Christoph Kaiser

High Quality Nb/Al-AlO$_x$/Nb Josephson Junctions

Technological Development and Macroscopic Quantum
Experiments

Karlsruher Schriftenreihe zur Supraleitung
Band 004

Herausgeber: Prof. Dr.-Ing. M. Noe
               Prof. Dr. rer. nat. M. Siegel

High Quality Nb/Al-AlO$_x$/Nb Josephson Junctions

Technological Development and Macroscopic Quantum Experiments

von
Christoph Kaiser

KIT Scientific Publishing

Dissertation, Karlsruher Institut für Technologie
Fakultät für Elektrotechnik und Informationstechnik, 2011
Hauptreferent: Prof. Dr. Michael Siegel
Korreferent: Prof. Dr. Alexey V. Ustinov

Impressum

Karlsruher Institut für Technologie (KIT)
KIT Scientific Publishing
Straße am Forum 2
D-76131 Karlsruhe
www.ksp.kit.edu

KIT – Universität des Landes Baden-Württemberg und nationales
Forschungszentrum in der Helmholtz-Gemeinschaft

KIT Scientific Publishing 2011
Print on Demand

ISSN 1869-1765
ISBN 978-3-86644-651-9

High Quality Nb/Al-AlO$_x$/Nb Josephson Junctions: Technological Development and Macroscopic Quantum Experiments

Zur Erlangung des akademischen Grades eines

DOKTOR-INGENIEURS

von der Fakultät für

Elektrotechnik und Informationstechnik

des Karlsruher Instituts für Technologie (KIT)

genehmigte

DISSERTATION

von

Diplom-Physiker Christoph Kaiser

geboren in Hameln

Tag der mündlichen Prüfung: 03. Februar 2011

Hauptreferent: Prof. Dr. Michael Siegel

Korreferent: Prof. Dr. Alexey V. Ustinov

Preface

This dissertation is the result of my work at the *Institut für Mikro- und Nanoelektronische Systeme (IMS)*, first at the *Universität Karlsruhe (TH)* and later at the *Karlsruher Institut für Technologie (KIT)*, which is the merger of the former with the *Forschungszentrum Karlsruhe (FZK)*. After all, this thesis would not have been possible without the help of a number of people, who I want to thank in the following (in German). After that, details about the way citations are given within this thesis are discussed.

Danksagung

Als erstes möchte ich Prof. Michael Siegel dafür sehr herzlich danken, dass er es mir ermöglicht hat, an diesem spannenden Thema zu arbeiten und meine Ergebnisse immer wieder auf internationalen Konferenzen vorstellen zu können. Ich habe das in mich gesetzte Vertrauen immer zu schätzen gewusst und das selbstbestimmte Arbeiten genossen. Auch für die unter seiner Führung erfolgreich abgeschlossene Promotion bin ich zu Dank verpflichtet.

Ein großer Dank geht auch an Prof. Alexey Ustinov für die Übernahme des Korreferats, viele interessante fachliche Diskussionen und die harmonische Zusammenarbeit bei verschiedenen Forschungsprojekten.

Eine ganz besondere Freude war für mich immer die Arbeit mit Dr. Roland Schäfer, von dem ich sowohl über experimentelle Methoden als auch über Festkörperphysik viel gelernt habe. Die leidenschaftlichen Diskussionen über Dekohärenz, Quantenmechanik und andere Themen werden mir fehlen! Und wer weiß, vielleicht werde ich sogar irgendwann wehmütig an die Suche nach Störquellen im Mischkryostaten zurückdenken? Für die schöne und erfolgreiche Zeit inklusive der Messung von Makroskopischem Quantentunneln und Subgap-Leckströmen noch einmal auf diesem Wege vielen Dank!

Danken möchte ich auch Dr. Thilo Bauch und Prof. Floriana Lombardi für die Messungen der *LC*-geshunteten Josephson-Kontakte und die Ermöglichung meines interessanten Forschungsaufenthaltes an der Chalmers University in Göteborg. Dabei ist auch meine Wertschätzung für die persische Küche und Sushi deutlich gestiegen.

Für sehr lehrreiche und spannende Messungen an Phase-Qubits bin ich Dr. Jürgen Lisenfeld sehr dankbar. Mir hat die gemeinsame Zeit im kalten Keller des CFN-Gebäudes viel Spaß gemacht und ich habe endlich die Bloch-Kugel inklusive *rotating frame* richtig verstanden. Ich hoffe, bei Jürgen ist mittlerweile Erholung von dem durch mich erzwungenen frühen Aufstehen eingetreten.

Ein ganz wesentlicher Baustein für das Zustandekommen dieser Arbeit waren natürlich die Mitarbeiter des IMS, die mir immer mit Rat und Tat zur Seite standen. Beginnen möchte ich hier mit Dr. Stefan Wünsch, der trotz einer anfänglich beträchtlichen Fehlanpassung meiner Physiker-Fragen und seiner Ingenieurs-Antworten nie die Geduld mit mir verloren hat. Dass das früher Undenkbare eingetreten ist und sich bei mir mittlerweile tatsächlich rudimentäre Kenntnisse in Hochfrequenztechnik entwickelt haben, ist allein sein Verdienst.

Weiterhin bin ich Dr. Kostya Ilin zu Dank verpflichtet, der mich insbesondere in die Kunst der Elektronenstrahllithographie eingeweiht hat. Auch bei Fragen rund um technologische Prozesse oder Anlagen hat er mich immer an seinem Erfahrungsschatz teilhaben lassen. Gleiches gilt für Dr. Manfred Neuhaus, der mir während meiner ersten Monate am IMS kurz vor seinem Ruhestand noch den ein oder anderen technologischen Kniff weitergeben konnte.

Einführungen in verschiedene Vakuumanlagen und die Fotolithographie habe ich auch von Alexander Stassen und Hansjürgen Wermund erhalten. Weiterhin standen mir beide bei vielen Problemen in der Prozessierung immer hilfreich und kompetent zur Seite, wofür ich ihnen herzlich danke. Insbesondere die liebevolle Pflege des technologischen Kernstücks dieser Arbeit, der UTS 500, durch Herrn Wermund sollte hier Erwähnung finden. Auch den zwei weiteren technischen Mitarbeitern am IMS, Karlheinz Gutbrod und Frank Ruhnau, möchte ich meinen Dank aussprechen. Herr Gutbrod hat mir vor allem durch die Herstellung meines hochkomplizierten Probenhalters sehr geholfen, während auf Frank bei Computerproblemen und der Betreuung des PSoC-Praktikums Verlass war.

Ein herzlicher Dank gilt Max Meckbach, der mich aus meiner Einsamkeit in der Josephson-Kontakt-Forschung am IMS erlöst hat. Dass wir in der Folge technologische Probleme gemeinsam oder abwechselnd lösen konnten, hat nicht nur mehr Spaß gemacht als zuvor, sondern auch die Fertigstellung dieser Arbeit erheblich beschleunigt. Ich hatte eine schöne Zeit mit Max, sowohl am Institut als auch in der Freizeit beim Tennisspielen oder Skifahren. Lass dich von den Junctions (und vor allem den Anlagen) nicht unterkriegen!

Meinen beiden weiteren Zimmergenossen, Alexander Scheuring und Axel Stockhausen, möchte ich für die lustige Zeit zusammen danken. Durch unsere Gespräche war immer ein sofortiger Frustabbau, unsinnige Blödelei oder ganz selten auch mal ein fachlicher Austausch möglich. Weiter so! Außerdem bin ich allen weiteren Mitarbeitern des IMS dafür dankbar, dass stets so eine nette Arbeitsatmosphäre herrschte. Grillen oder Fußballspiele in der Mittagspause haben das Doktorandenleben immer wieder versüßt. Namentlich danke ich Erich Crocoll, Doris Duffner, Gerd Hammer, Dagmar Rall, Matthias Hofherr und Petra Probst.

Sehr wesentlich zum Gelingen dieser Arbeit haben auch die Studenten beigetragen, die im Rahmen von Praktika, HiWi-Stellen und Studien- oder Diplomarbeiten Teilthemen bearbeitet haben. Dementsprechend danke ich Andreas Graf, Daniel Bruch, Martin Berov, Matthias Birk, Franz Schuster, Maher Rezem, Christian Goebel, Jochen Antes, Max Bauer, Michael Merker und Sebastian Skacel für die angenehme Zusammenarbeit.

Mein größter Dank gilt meiner Frau Sierra, die mir in vielen frustrierenden, anstrengenden und hoffnungslos scheinenden Zeiten Unterstützung, Liebe und Lebensmut gegeben hat. Danke, dass ich dich habe und du so bist, wie du bist!

Citations in this Thesis

To facilitate the distinction of references for the reader, I decided to group the cited publications in three classes, which are represented by different abbreviations:

- External publications are given by numbers in square brackets, sorted by their first occurrence in the text. Example:

[1]   B. D. Josephson. *Possible New Effects in Superconductive Tunnelling.* Physics Letters, **1**, 251 (1962).

- My own publications in scientific journals are given in a separate list. They are cited as the first letters of the last names of the first $3-4$ authors, followed by the year of publication, all in square brackets. A "+"-sign indicates more than 4 authors. Example:

[KSW+10]   Ch. Kaiser, S. T. Skacel, S. Wünsch, R. Dolata, B. Mackrodt, A. Zorin and M. Siegel. *Measurement of dielectric losses in amorphous thin films at gigahertz frequencies using superconducting resonators.* Superconductor Science and Technology, **23**, 075008 (2010).

- Student theses, which I supervised during my PhD thesis, make the third list. They are referred to by the first three letters of the last name of the student, followed by the year of completion. Example:

[Ant09]   Jochen Antes. *Entwurf und Konstruktion einer Steuerung für die kontrollierte Oxidation von Aluminium.* Studienarbeit, Institut für Mikro- und Nanoelektronische Systeme, Universität Karlsruhe (TH) (2009).

The three complete lists are given at the end of the thesis, as indicated in the table of contents.

Zusammenfassung

Diese Arbeit beschreibt die Entwicklung eines technologischen Gesamtprozesses zur Herstellung hochqualitativer Nb/Al-AlO$_x$/Nb Josephson-Kontakte und die Durchführung makroskopischer Quantenexperimente mit diesen Bauelementen.

Josephson-Kontakte sind Tunnelkontakte mit der Schichtfolge Supraleiter-Isolator-Supraleiter (SIS), deren Funktionsweise darauf beruht, dass die Phase der supraleitenden quantenmechanischen Wellenfunktion in ihnen einen definierten Sprung aufweist. Diese Phasenänderung kann durch die Josephson-Gleichungen mit dem Strom und der Spannung über den SIS-Kontakt in Beziehung gesetzt werden. Josephson-Kontakte werden heute bereits in vielen Anwendungen eingesetzt, beispielsweise zur Definition der SI-Einheit Volt, in ultraschneller Elektronik (RSFQ), als Strahlungsdetektoren oder in SQUIDs, welche die empfindlichsten Sensoren für magnetischen Fluss sind. Ihre Herstellung erfolgt dabei standardmäßig in der Nb/Al-AlO$_x$/Nb-Technologie, die durch hohe Ausbeute und kleine Parameterstreuung die Fabrikation integrierter Schaltungen mit zehntausenden Josephson-Kontakten erlaubt.

Da der Phasensprung über einen SIS-Kontakt ein makroskopischer, quantenmechanischer Freiheitsgrad ist, werden Josephson-Kontakte seit den 1980er Jahren als Modellsysteme für die Erforschung makroskopischer Quantenphänomene eingesetzt. Ein Beispiel hierfür ist das Makroskopische Quantentunneln (MQT), bei dem die Phasensprung-Variable durch eine Potentialbarriere tunnelt. In den letzten Jahren wurden Josephson-Kontakte auch zur Herstellung von supraleitenden Quantenbits (Qubits), welche die Bausteine von Quantencomputern darstellen, verwendet. Diese Qubits haben gegenüber anderen Qubit-Architekturen (wie z.B. Ionen, Kernspins etc.) den Vorteil, dass sie in Chiptechnologie hergestellt werden können und deshalb leicht zu skalieren sind, sowie dass ihre Parameter frei einstellbar sind. Ein Nachteil ist allerdings, dass die bislang hergestellten supraleitenden Qubits eine beschränkte Kohärenzzeit aufweisen, d.h. dass die kohärente Evolution ihres Quantenzustandes schnell abklingt. Aus technologischer Sicht reduzieren beispielsweise signifikante Leckströme in den Josephson-Kontakten oder Tunnelsysteme (TLS) in amorphen Dielektrika, die in den Schaltkreisen verwendet werden, die Kohärenzzeit. Letzteres ist insbesondere bei den im Rahmen dieser Arbeit untersuchten *Phase Qubits*, die aus einem Josephson-Kontakt in einem supraleitenden Ring bestehen, ein Problem. Dadurch ergibt sich die ingenieurstechnische Aufgabe, den SIS-Fabrikationsprozess zu optimieren sowie die verwendeten Materialien zu untersuchen und zu verbessern, damit Josephson-Kontakte ausreichender Qualität zur Durchführung makroskopischer Quantenexperimente und insbesondere für supraleitende Qubits mit längeren Kohärenzzeiten hergestellt werden können.

Das Ziel der vorliegenden Arbeit bestand darin, den Fabrikationsprozess für Nb/Al-AlO$_x$/Nb-Josephson-Kontakte am IMS weiterzuentwickeln sowie die Kontaktabmessungen zu miniaturisieren, und mit den daraus gewonnenen Proben makroskopische Quantenexperimente zu konzipieren und durchzuführen. Dazu wurden zuerst die durch den Herstellungsprozess erzielte Kontaktqualität analysiert und potentielle Verbesserungsmöglichkeiten identifiziert. Darauf basierend wurde der Fabrikationsprozess optimiert, beispielsweise durch Änderung der Prozessreihenfolge, Reduzierung der Filmspannung in Nb-Filmen und Einstellen eines neuen Arbeitspunktes beim reaktiven Ionen-Ätzen. Zusätzlich wurde ein neuer Herstellungsprozess für Josephson-Kontakte mit Abmessungen im Submikrometer-Bereich entwickelt. Dies erfolgte durch den Einsatz von Elektronenstrahllithographie und ihre detaillierte Optimierung sowie durch die Entwicklung des neuartigen Konzeptes einer Aluminium-Maske für die Kontaktdefinition. Die Charakterisierung der so hergestellten Josephson-Kontakte ergab eine exzellente Qualität sowie eine ausreichende Skalierbarkeit für die angestrebten Anwendungen.

Für die meisten Anwendungen von Josephson-Kontakten, aber insbesondere für supraleitende Qubits, ist es wichtig, dass die für die benötigten Isolationsebenen eingesetzten Materialien kleine dielektrische Verluste aufweisen. Aus diesem Grund wurde in dieser Arbeit eine Methode zur direkten und quantitativen Untersuchung dielektrischer Verluste in amorphen Dünnfilmen entwickelt und zur Messung solcher Verluste in Materialien verwendet, die typischerweise in der SIS-Fabrikation zum Einsatz kommen. Dadurch konnte entschieden werden, dass als Isolatorschicht thermisch aufgedampftes SiO besser für die angestrebten Experimente geeignet ist als das durch RF-Sputtern ebenfalls am IMS herstellbare SiO_2. Weiterhin wurde mit der neu entwickelten Methode die Frequenzabhängigkeit der dielektrischen Verluste in verschiedenen Materialien bestimmt, wodurch festgestellt werden konnte, dass die für die Verluste in diesen glasartigen Materialien verantwortlichen Dipole in Vielteilchenwechselwirkung miteinander stehen.

Als erstes makroskopisches Quantenexperiment wurde ein MQT-Experiment durchgeführt, wobei die Übergangstemperatur vom thermischen zum Quantenregime auf ihre Abhängigkeit von der Kontaktgröße und von einem externen Magnetfeld untersucht wurde. Beide Sachverhalte waren zuvor noch nie experimentell erforscht worden. Es zeigte sich, dass sowohl die jeweiligen Abhängigkeiten sehr gut mit den theoretischen Erwartungen übereinstimmten, als auch die Quanteneigenschaften der Proben mit großer Genauigkeit aus den Designparametern vorhersagbar waren. Dies zeigt, dass mit dem neu entwickelten Fabrikationsprozess hergestellte Josephson-Kontakte gut für makroskopische Quantenexperimente geeignet sind. Weiterhin wurde festgestellt, dass die Quantentunnelrate durch die Potentialbarriere bei bestimmten Werten des magnetischen Flusses signifikant zunahm. Dies ist ein gänzlich unbekannter Effekt, der vermutlich durch die Ausbildung von Eigenresonanzen in den Kontakten erklärt werden kann.

In einem weiteren Experiment wurde die Quantendynamik eines Schaltkreises aus einem Josephson-Kontakt, einer Induktivität und einem Kondensator untersucht. Die theoretische Erwartung hierbei ist, dass sich ein zweidimensionales Potential ausbildet, welches zu zwei neuen Energieskalen im System führt. Dies konnte durch Mikrowellen-Spektroskopie bestätigt werden, was belegt, dass sich dieses komplexe und $200 \times 650 \ \mu m^2$ große System tatsächlich wie ein einziges Quantenobjekt verhält. Dieses Ergebnis ist sowohl wichtig für das Verständnis des Quantenverhaltens von Josephson-Kontakten aus Hochtemperatur-Supraleitern, da bei ihrer Herstellung parasitäre Induktivitäten und Kapazitäten teilweise unvermeidbar sind, als auch für das Design und die Operation supraleitender Phase Qubits, weil hier oft zur Verschiebung des Arbeitspunktes Kondensatoren durch induktive Leitungen an das Qubit angeschlossen werden.

Abschließend wurde die entwickelte SIS-Technologie verwendet, um Phase Qubits herzustellen und zu vermessen. Hier wurde die kohärente Überlagerung der zwei Qubit-Zustände nachgewiesen, was das wesentliche Kriterium dafür darstellt, dass es sich tatsächlich um ein makroskopisches Quantensystem handelt. Weiterhin konnte durch charakteristische Messungen gezeigt werden, dass sich das Qubit entsprechend der theoretischen Vorhersagen verhielt. Ein Vergleich mit allen anderen, in ähnlichem Design und mit Nb/Al-AlO$_x$/Nb-Technologie hergestellten Phase Qubits aus der Literatur ergab, dass die im Rahmen dieser Arbeit gemessenen Kohärenzzeiten etwas länger waren. Ein möglicher Grund hierfür könnte in den sehr niedrigen Leckströmen der Josephson-Kontakte liegen, sowie in der neu entwickelten Aluminium-Maske, da sie zusätzliche chemische Prozessschritte vermeidet. Für den Einsatz in Quantencomputern müsste die Kohärenzzeit aber noch deutlich verlängert werden. Ansätze hierfür werden in dieser Arbeit diskutiert und wurden teilweise auch schon experimentell vorbereitet.

Zusammengefasst kann gesagt werden, dass die im Rahmen dieser Arbeit entwickelte Technologie bereits erfolgreich für verschiedene makroskopische Quantenexperimente eingesetzt wurde, wodurch einige ungeklärte theoretische Fragestellungen beantwortet werden konnten. Zudem sind viele interessante Experimente für die Zukunft denkbar, insbesondere dadurch, dass Phase Qubits mit außergewöhnlich langen Kohärenzzeiten herstellbar sind.

Table of Contents

1 Introduction

Since Brian D. Josephson predicted the possibility of superconductive tunneling through an insulating barrier between two superconducting electrodes [1], Josephson junctions have become a tool vastly used in science and engineering. Important technical applications include ultrafast electronics (RSFQ = Rapid Single-Flux-Quantum) [2], the definition of the SI[1] unit volt by the Josephson voltage standard [3, 4], radiation sources [5], SIS receivers [6], digital-to-analogue converters [7] and the most sensitive detectors for magnetic flux, SQUIDs [8]. Since a Josephson voltage standard for example involves tens of thousands of junctions, fabrication facilities all over the world have put their focus on a scalable technology with high yield and small parameter spreads. As a standard technology, $Nb/AlO_x/Nb$ Josephson junctions have been established, in which a supercurrent tunnels through the insulating aluminum oxide (AlO_x) barrier from one superconducting niobium (Nb) electrode to the other. Today, the $Nb/AlO_x/Nb$ technology has proven to be very mature and indeed fulfill the requirements described above, so that the mentioned devices are frequently and reliably used.

In recent years, however, a new field of application and research has arisen, which explicitly takes advantage of the quantum nature of superconductivity. The charge carriers in superconductors are Cooper pairs, which have zero spin and hence have a bosonic character. Consequently, the superconducting state can be described by one macroscopic, quantum mechanical wavefunction, similar to a Bose-Einstein condensate. In Josephson junctions, the phase of this wavefunction exhibits a non-continuous jump, which can be related to the current and the voltage across the junction. Hence Josephson junctions allow direct manipulation and analysis of the quantum mechanical, macroscopic superconducting wavefunction.

This is the reason why superconducting Josephson circuits are considered to be the model system to answer a scientific question that has troubled physicists for nearly a century: While it has become clear that quantum theory describes nature on the atomic scale very accurately, it is also clear that our macroscopic, everyday world does not exhibit phenomena like entanglement or the superposition principle, which allows a system to be in multiple distinct states simultaneously. In his famous "Cat Paradox" published in 1935, Erwin Schrödinger pointed out which peculiar effects it would have if the laws of quantum mechanics were directly transferred to our everyday world [9]. In his gedanken experiment, a closed box contains a radioactive atom, whose decay is linked to the health of a cat, so that until somebody opens the box, the cat will be in a superposition state of dead and alive. Since such macroscopic manifestations of quantum mechanics are unimaginable and simply unobserved, it had always remained unclear how and why the crossover from the microscopic quantum world to the macroscopic classical world takes place.

---

[1]International System of Units, abbreviated SI from the French *le Système international d'unités*

As mentioned above, superconducting circuits involving Josephson junctions can help to shed light on this question. Since they are macroscopic systems containing some 10^{23} particles but can be described by one single quantum mechanical wave function, there was large interest to see whether they would indeed behave quantum mechanically. A first answer was given when the tunneling of the phase difference over a Josephson junction (which is a macroscopic degree of freedom) through a potential barrier was observed in 1981 [10]. This effect is analogous to the radioactive decay of an atom and is called macroscopic quantum tunneling (MQT). However, as A. L. Leggett pointed out in 1985 [11], an experiment that would indeed reconstruct the case described by Schrödinger should involve the superposition of two macroscopically distinct states like the clockwise and counter-clockwise circulating currents in a SQUID. (A SQUID is a superconducting ring involving one (RF-SQUID) or two (DC-SQUID) Josephson junctions.) This was finally demonstrated in the year 2000 in a superconducting ring interrupted by a DC-SQUID [12]. Already one year earlier, the same superposition effect had been demonstrated on junctions operated in the charge regime [13]. Later, similar experiments were carried out on a current-biased Josephson junction [14], a superconducting loop containing three Josephson junctions [15] and hybrid flux-charge systems [16]. So such Schrödinger cat states do indeed exist, but it was also found that they decay into classical states with a coherence time that is mostly far shorter than one µs.

The experiments mentioned above investigated superpositions of two distinct states and were hence carried out on circuits which were designed to be quantum mechanical two level systems. Since these systems fulfill all criteria needed for the implementation of a quantum computer [17], they can be used as quantum bits (qubits). Quantum computers can solve certain problems exponentially faster than any classical computer, such as the simulation of quantum systems [18], the prime factorization of large numbers [19] or searching an unsorted database [20]. Josephson junction qubits offer a multitude of advantages over typical microscopic qubits such as nuclear spins, ions, etc.: They can be fabricated on chip and hence integrated with classical electronics, they can be scaled up relatively easily, they can be fabricated with common thin film technology and their properties can be designed at will. This is why a large interest in Josephson quantum circuits has developed in the past ten years.

The main problems of superconducting qubits are their limited coherence times, since the implementation of a real quantum computer only makes sense if a sufficient time for calculations is available. The reasons for the short coherence times may vary for different types of superconducting qubits and are still the subject to a large research effort. For phase qubits (a superconducting ring interrupted by one Josephson junction), which will be investigated in this thesis, it has been found that their coherence times are significantly longer for circuits involving $Al/AlO_x/Al$ junctions than for those based on $Nb/AlO_x/Nb$ junctions [21, 22]. This is unfortunate, since the latter is the standard technology in all Josephson junction foundries and offers all the advantages of scalability, high yield and small parameter spreads described above. Furthermore, it was found that the material used for insulating layers in the qubit strongly influences the coherence times [23]. Both effects are attributed to spurious two-level systems (TLS), which couple to the qubit and extract the quantum information from it. Macroscopically, these TLS manifest themselves as dielectric loss. One can say that while many physicists still use superconducting qubits to principally proof various quantum effects on a macroscopic scale, the question how the coherence times of these devices can be improved has turned into an engineering problem involving a large effort in materials research.

The goal of this thesis was to develop a technological process that would allow the fabrication of such $Nb/AlO_x/Nb$ based quantum devices. In order to understand the physics of such structures, a basic introduction to macroscopic quantum effects in Josephson junctions is given in chapter 2. Afterwards, quality parameters which are important for quantum experiments are discussed in chapter 3. As small Josephson junctions are favorable for most quantum experiments, the focus in development of the fabrication process (described in chapter 4) was put on a technology for small Josephson junctions of very high quality. This newly developed technological process was used to fabricate all Josephson junctions in this thesis. At first, the junction quality was thoroughly characterized in order to see whether the fabricated junctions could be used for quantum experiments (chapter 5). Furthermore, in order to investigate the dielectric losses in the employed insulating layers, a method was developed to reliably and directly measure dielectric losses in amorphous thin films. This method, which is described in chapter 6, allowed to investigate the frequency dependent losses in various insulating layers and should allow to reduce dielectric losses in the future in order to obtain qubits with longer coherence times. In the following, MQT measurements were carried out to show that the fabricated Josephson junction devices are indeed suitable for the observation of macroscopic quantum effects. Especially the dependence of MQT on the junction size and on magnetic field was investigated (chapter 7), as both effects had not been studied experimentally before. Additionally, the quantum dynamics of a Josephson junction shunted by an inductance and a capacitance were investigated (chapter 8) in order to understand such complex Josephson circuits, as they are often used in qubit design. To see whether the developed technology is able to produce quantum bits, superconducting phase qubits were fabricated and characterized (chapter 9). It was also a goal of this investigation to find out if the short coherence times in $Nb/AlO_x/Nb$ based phase qubits are inherent to the technology or can be improved.

Even if the goal of building a quantum computer would never be reached, a high-quality fabrication technology for Josephson junctions based quantum devices still remains important. As their properties can be designed at will, Josephson junctions have and will be used as model systems to study many quantum effects which are unattainable with natural quantum systems. Examples are quantum dynamics of intermediately damped and overdamped systems, the details of quantum tunneling and, of course, the quantum physics of macroscopic systems.

2 Macroscopic Quantum Phenomena in Josephson Junctions

This chapter aims at explaining the basic physics behind the Josephson junction experiments performed in this thesis. It begins by giving a brief introduction to the physics of superconductivity. Then, the dynamics of Josephson junctions are discussed in some detail and the phenomenon of macroscopic quantum tunneling is explained. Finally, superconducting phase qubits are briefly introduced and their operation scheme is discussed. A more detailed introduction to MQT and phase qubits is given in the corresponding chapters.

2.1 Superconductivity

In 1908, the Dutch physicist Heike Kammerlingh Onnes succeeded in the liquefaction of the noble gas helium. As the boiling point of He is at 4.2 K, this opened up a new temperature range for experiments. Subsequently, Onnes himself used this opportunity to study the electrical resistance of mercury at low temperatures. In 1911, he found that the electrical resistance of his sample dropped down to zero below 4.2 Kelvin. This was the discovery of superconductivity.

The superconducting state persists below a critical temperature T_c, a critical magnetic field H_c and a critical current density $j_{s,c}$. All these parameters are material dependent. In 1933, Fritz Walther Meissner and Robert Ochsenfeld discovered that a magnetic field is expelled from the inside of a superconductor, which, in combination with its further properties, means that superconductivity is a thermodynamic phase of its own. Two years later, the brothers Fritz and Heinz London found equations which describe the relation of the supercurrent density $\vec{j}_s$ to the electric field $\vec{E}$ and the magnetic flux density $\vec{B}$:

$$\mu_0 \lambda_L^2 \frac{\mathrm{d}\vec{j}_s}{\mathrm{d}t} = \vec{E} \tag{2.1}$$

$$\mu_0 \lambda_L^2 \vec{\nabla} \times \vec{j}_s = -\vec{B} \tag{2.2}$$

These equations are called the first and the second London equation, respectively. Here, μ_0 denotes the vacuum permeability and

$$\lambda_L = \sqrt{\frac{m_s}{\mu_0 q_s^2 n_s}} \tag{2.3}$$

is called the London penetration depth, with q_s being the charge, m_s the mass and n_s the density of the superconducting charge carriers. The meaning of λ_L becomes clear when

(2.2) is combined with Ampère's circuital law $\vec{\nabla} \times \vec{B} = \mu_0 \vec{j}_s$, so that we obtain

$$\vec{\nabla}^2 \vec{B} = \frac{1}{\lambda_L^2} \vec{B}. \tag{2.4}$$

Obviously, the magnetic field decays exponentially inside the superconductor with a spatial constant λ_L, which gives a more detailed picture of the Meissner effect. The field penetration is inhibited by screening currents flowing in a surface layer with thickness λ_L. It is important to note that experimental values of λ_L can differ significantly from the theoretical value (2.3).

It was not until 1957 that a microscopic theory explaining superconductivity, the BCS theory (named in reference to its founders John Bardeen, Leon Neil Cooper and John Robert Schrieffer), was developed. Already in 1950, Herbert Fröhlich had shown that an effective attractive force between two conduction electrons can evolve in solid state systems, mediated by the exchange of a virtual photon. This attractive force can lead to the formation of bound states of two electrons with opposing momentum and spin. These new composite particles, commonly known as Cooper pairs, have spin zero and hence obey the Bose-Einstein statistics, meaning that they all condense in the same quantum state at low temperatures. As a superconductor is a macroscopic object, we are dealing with a macroscopic quantum state. It can be described by a single quantum mechanical wavefunction

$$\psi = |\psi| \exp i\phi = \sqrt{n_s} \exp(i\phi), \tag{2.5}$$

where $\phi(\vec{x}, t)$ is the space and time dependent phase and the amplitude $|\psi(\vec{x}, t)|^2 = n_s$ is given by the Cooper pair density. It should be mentioned that already in 1950, Ginzburg and Landau had introduced a complex order parameter ψ in their thermodynamical theory, which is equivalent to the wavefunction in the BCS theory. Cooper pairs cannot be lifted in energy as a whole, so that the superconducting state can only be excited by breaking them up. These excitations are called quasiparticles and behave like single electrons in most aspects. In order to break up Cooper pairs, their binding energy 2Δ must be provided to the system, which separates the quasiparticle states energetically from the Cooper pair states by an energy gap Δ.

The size of the Cooper pairs is temperature independent and is given by the BCS coherence length ξ_0, which is much larger than the distance between individual Cooper pairs. This is an explanation for the fact that the superconducting wavefunction is phase coherent throughout a superconducting sample. The temperature dependent Ginzburg-Landau coherence length $\xi(T)$ is the distance on which ψ can vary. In the dirty limit, $\xi(T)$ is given by [24]

$$\xi(T) = 0.855 \frac{\sqrt{\xi_0 \ell}}{\sqrt{1 - T/T_c}}, \tag{2.6}$$

where T denotes the temperature and ℓ the electronic mean free path.

2.2 Josephson Junctions

Quantum mechanics predict that potential barriers which cannot be overcome according to the laws of classical physics can be penetrated by the tunneling effect. In 1962, Brian David

Josephson showed that such a tunneling effect can also be expected for two weakly coupled superconductors [1]. The most common way of realizing such a weak coupling is to put a thin insulating layer between two superconducting electrodes. Such superconductor-insulator-superconductor structures are commonly known as SIS Josephson junctions. More detailed introductions to the physics of Josephson junctions can be found in [24–26].

2.2.1 Josephson Equations

Josephson's analysis is especially noteworthy as it predicts the occurrence of a tunneling current even in absence of a voltage drop, i.e. the formation of a supercurrent through the insulating barrier. This supercurrent will be limited to a critical current density j_c, which is orders of magnitude smaller than the critical current density $j_{s,c}$ in the bulk superconducting electrodes. For the case of two equal superconducting electrodes, so that $n_{s1} = n_{s2}$, this is expressed by the first Josephson equation

$$j = j_c \sin \varphi.$$ (2.7)

Here, the gauge-invariant phase difference across the junction is given by

$$\varphi = \phi_2 - \phi_1 - \frac{2e}{\hbar} \int_1^2 \vec{A} \mathrm{d}\vec{l},$$ (2.8)

where e is the elementary charge, $\hbar = h/(2\pi)$ is the reduced Planck constant, $\vec{A}$ is the magnetic vector potential and $\mathrm{d}\vec{l}$ is the line integral. If the bias current I exceeds the critical current of the junction $I_c = j_c \cdot A$ (with A being its area), an additional quasiparticle current is needed for charge transport, so that a voltage will form across the junction. The second Josephson equation relates this voltage V to the time evolution of the phase difference:

$$\frac{\mathrm{d}\varphi}{\mathrm{d}t} = \frac{2e}{\hbar} \cdot V = \frac{2\pi}{\Phi_0} \cdot V,$$ (2.9)

where $\Phi_0 = h/2e$ is the magnetic flux quantum (see Appendix A).

The combination of (2.7) and (2.9) shows that applying a voltage across the junction leads to an alternating current

$$I = I_c \sin \left(\frac{2e}{\hbar} V \cdot t + \varphi_0 \right),$$ (2.10)

which is oscillating with the Josephson frequency

$$f_J = \frac{1}{\Phi_0} V = 483.6 \frac{\mathrm{GHz}}{\mathrm{mV}} \cdot V.$$ (2.11)

The occurrence of an AC current implies the emission of photons. Inversely, if microwave photons are radiated onto a junction, quantized voltage steps (Shapiro steps) will form. This effect is commonly known as inverse Josephson effect and has been the basis for the definition of the physical unit volt since 1990.

Since a change in the supercurrent and hence the Josephson phase will lead to a nonzero voltage according to (2.9), a Josephson junction acts as a nonlinear inductance. This becomes clear if (2.7) is differentiated and combined with (2.9), so that

$$L_{\mathrm{J}} = V \left(\frac{\mathrm{d}I}{\mathrm{d}t}\right)^{-1} = \frac{\hbar}{2e}\frac{1}{I_{\mathrm{c}}\cos\varphi} = L_{\mathrm{J0}}\left[1-\left(\frac{I}{I_{\mathrm{c}}}\right)^2\right]^{-1/2} \tag{2.12}$$

is obtained as the Josephson inductance. Here $L_{\mathrm{J0}} = \hbar/(2eI_{\mathrm{c}})$, while the last transformation simply uses (2.7) and $\sin^2\varphi + \cos^2\varphi = 1$.

2.2.2 RCSJ Model

The dynamics of a Josephson junction are usually described by the RCSJ (resistively and capacitively shunted junction) model introduced by W. C. Stewart [27] and D. E. McCumber [28]. This model applies to short junctions, for which the Josephson phase φ can be considered a point-like variable. This is the case if the dimensions of the junction are smaller than the Josephson penetration depth [29]

$$\lambda_{\mathrm{J}} = \sqrt{\frac{\Phi_0}{2\pi\mu_0 j_{\mathrm{c}}(2\lambda_{\mathrm{L}}+t_{\mathrm{ox}})}}\,, \tag{2.13}$$

where μ_0 is the vacuum permeability and t_{ox} is the thickness of the insulating barrier. Typical values for λ_{J} are in the range of 5 to 30 µm. All junctions investigated within this thesis have dimensions clearly below λ_{J}, so that they can be described by the RCSJ model. The latter takes into account that in parallel to the Josephson current described by (2.7), the junction will also exhibit a capacitance C due to its plate capacitor geometry as well as a resistance R for quasi-particle transport (see Figure 2.1). Hence, the bias current I is composed of

$$I = I_{\mathrm{c}}\sin\varphi + \frac{V}{R} + C\frac{\partial V}{\partial t}\,, \tag{2.14}$$

which, by the use of (2.9), can be transformed into the more revealing form

$$I = I_{\mathrm{c}}\sin\varphi + \frac{1}{R}\frac{\Phi_0}{2\pi}\dot\varphi + C\frac{\Phi_0}{2\pi}\ddot\varphi\,. \tag{2.15}$$

This is the equation of motion for a Josephson junction. Instead of solving this differential equation of second order, we can employ the equivalent picture of a particle of mass

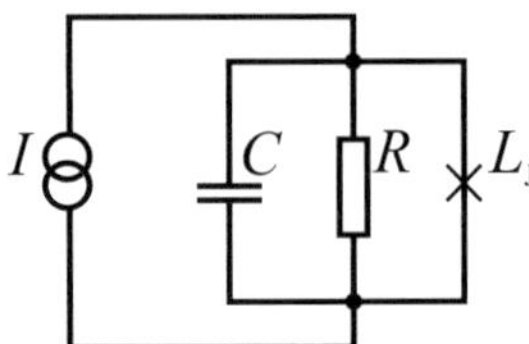

Figure 2.1: Resistively and capacitively shunted junction (RCSJ) model.

$M = C(\Phi_0/2\pi)^2$ exposed to friction with coefficient $\eta = 1/R \cdot (\Phi_0/2\pi)^2$ and moving along φ in a potential $U(\varphi)$, so that we obtain

$$M\ddot{\varphi} + \eta\dot{\varphi} + \frac{\partial U(\varphi)}{\partial\varphi} = 0. \tag{2.16}$$

Here, the potential is given by

$$U(\varphi) = E_{\mathrm{J}}\left(1 - \cos\varphi - \gamma\varphi\right), \tag{2.17}$$

where $\gamma = I/I_{\mathrm{c}}$ denotes the normalized bias current while $E_{\mathrm{J}} = \Phi_0 I_{\mathrm{c}}/2\pi$ is called the Josephson coupling energy. This potential is shown in Figure 2.2 and is often referred to as the tilted washboard potential. We see that for $\gamma < 1$, if no thermal or quantum fluctuations are present, the particle will be trapped behind a potential barrier

$$\Delta U = 2E_{\mathrm{J}}\left(\sqrt{1 - \gamma^2} - \gamma\arccos\gamma\right), \tag{2.18}$$

which means that $\langle\dot{\varphi}\rangle = 0$ and hence, according to (2.9), no average voltage across the junction is present (this is called the zero-voltage state). In the potential well, the phase oscillates with the bias current dependent plasma frequency, which is simply the eigenfrequency of the junction's *LC* circuit:

$$\omega_{\mathrm{p}} = \frac{1}{\sqrt{L_{\mathrm{J}}C}} = \sqrt{\frac{2eI_{\mathrm{c}}}{\hbar C}}\left(1 - \gamma^2\right)^{1/4} = \omega_{\mathrm{p}0}\left(1 - \gamma^2\right)^{1/4}. \tag{2.19}$$

If the bias current exceeds the critical value $\gamma = 1$, the potential barrier ΔU disappears, so that the particle starts rolling down the potential. It attains a nonzero average velocity

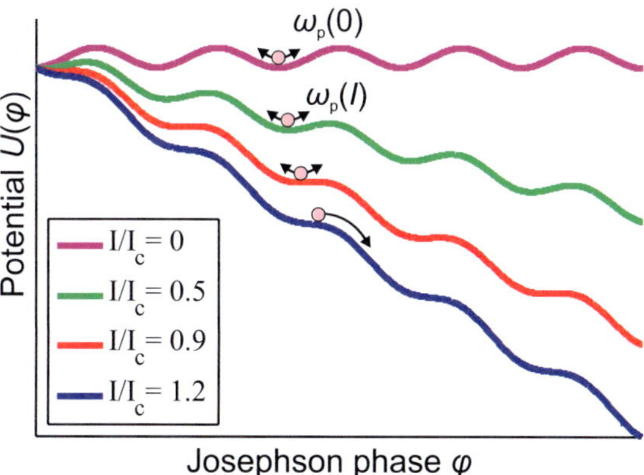

Figure 2.2: The dynamics of a Josephson junction can be described by a particle moving along a coordinate φ in a potential (2.17). This potential $U(\varphi)$ is plotted here for different values of the normalized bias current $\gamma = I/I_{\mathrm{c}}$. Due to its form, $U(\varphi)$ is commonly known as the tilted washboard potential.

$\langle \dot{\varphi} \rangle \neq 0$, which according to (2.9) is equivalent to a voltage drop over the junction. Consequently, this situation is called the voltage state or the running state. When the bias current is now successively changed back below I_c and down to zero, a potential barrier will first form and then gradually increase, so that the particle will eventually stop its movement and go back to the zero-voltage state. Whether this happens right below $\gamma = 1$ or only close to $\gamma = 0$ depends on the relation of the friction η to the particle mass M. A very clear and common measure for the damping in a Josephson junction can be found by using the normalized time $\tilde{t} = 2eI_cR/\hbar \cdot t$ and transforming equation (2.15) into the dimensionless form

$$\beta_C \, \ddot{\varphi} + \dot{\varphi} + \sin\varphi - \gamma = 0. \tag{2.20}$$

This introduces the Stewart-McCumber parameter β_C, which is given by

$$\beta_C = \frac{2e}{\hbar} I_c R^2 C. \tag{2.21}$$

For $\beta_C \gg 1$, the junction is in the underdamped regime, so that the particle will continue rolling until $\gamma \approx 0$ is reached. This is equivalent to a hysteresis in the *IV* curve, as will be discussed in more detail in section 3.1. Inversely, for $\beta_C \ll 1$ the particle will stop moving right after the occurrence of the first potential barrier at $\gamma \approx 1$, so that no hysteresis in the *IV* curve can be observed. This is called the overdamped regime.

2.2.3 Macroscopic Quantum Tunneling

Due to the presence of thermal or quantum fluctuations in real experiments, the underdamped junction will always switch to the voltage state before the theoretical critical current I_c is reached. This phenomenon is called premature switching and the observed critical current is commonly known as the switching current $I_{sw} \lesssim I_c$. The difference between I_{sw} and I_c is rather small and insignificant for many experiments as well as for the quality parameters discussed in chapter 3. Hence, no distinction is made between them in literature and it is just spoken of a critical current I_c. The same will be done in this thesis. Only if the difference between I_{sw} and I_c is crucial and shall be emphasized, the theoretical critical current will be denoted as I_{c0}. In this section, the difference is the crucial point: premature switching means that the junctions switches to the voltage state at a critical current $I_c < I_{c0}$.

At finite temperatures, the thermal energy k_BT (k_B being Boltzmann's constant) can lift the particle over the potential barrier before the critical current $\gamma = 1$ is reached, so that it will start rolling down the potential. The thermal escape from the potential well occurs with a rate [30, 31]

$$\Gamma_{th} = a_t \frac{\omega_p}{2\pi} \exp\left(-\frac{\Delta U}{k_B T}\right), \tag{2.22}$$

where a_t is a temperature and damping dependent prefactor, which will be discussed in more detail in section 7.2.2. For $T \to 0$, where $\Gamma_{th} \to 0$, premature switching will still be present due to quantum tunneling through the potential barrier. As the phase difference over the Josephson junction is a macroscopic variable, this phenomenon is often referred to as macroscopic quantum tunneling (MQT). We can write the quantum tunneling rate as [32–35]

$$\Gamma_{\mathrm{q}} = a_{\mathrm{q}} \frac{\omega_{\mathrm{p}}}{2\pi} \exp(-B)\,, \tag{2.23}$$

where $a_{\mathrm{q}} = \sqrt{864\pi\Delta U/\hbar\omega_{\mathrm{p}}}\exp\left(1.430/Q\right)$ and $B = 36\Delta U/(5\hbar\omega_{\mathrm{p}})\cdot(1+0.87/Q)$ are temperature independent parameters. $Q = \omega_{\mathrm{p}}RC$ describes the damping the junction and will be discussed in detail in section 7.2.2. This means that by measuring the switching events of a junction for decreasing temperature, one will see a temperature dependent behavior until a crossover to the temperature independent quantum regime is observed. The crossover temperature T_{cr} is approximately given by [33, 36]

$$T_{\mathrm{cr}} = \frac{\hbar\omega_{\mathrm{p}}}{2\pi k_{\mathrm{B}}}\,. \tag{2.24}$$

In the limit of large Q, the escape rate is expected to approach the temperature independent expression (2.23) quickly [34, 37, 38] once the temperature falls below T_{cr}, which means that the crossover should be clearly observable. It can be nicely visualized by measuring the escape rate and assuming that (2.22) is valid for all temperatures. In this way, one obtains a virtual escape temperature T_{esc} that can be compared to the actual bath temperature T. In the thermal regime, one should obtain $T_{\mathrm{esc}} = T$ while in the quantum regime, one should get $T_{\mathrm{esc}} = T_{\mathrm{cr}} = \mathrm{const}$.

More theoretical and experimental details of MQT, particularly concerning the influence of junction size, damping and an external magnetic field, will be discussed in chapter 7.

2.2.4 Resonant Activation

In the precedent section, we have employed the semi-classical picture of a phase particle which may tunnel through the potential barrier. In the full quantum mechanical picture, it needs to be taken into account that the energy levels in the potential well will be quantized.

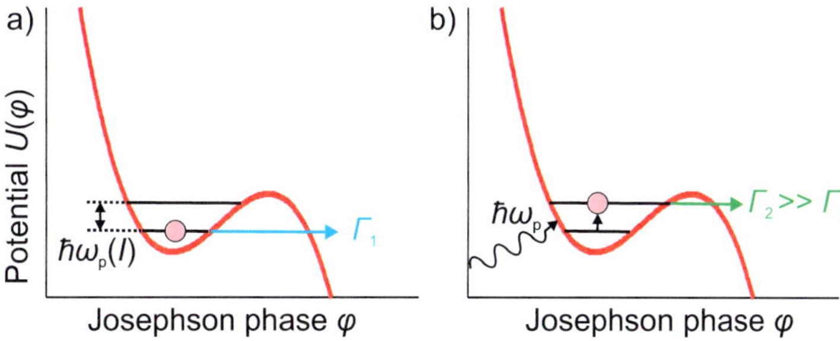

Figure 2.3: a) In the full quantum mechanical picture, the energy in the potential well is quantized. b) Irradiation of the junction with resonant microwave photons lifts the phase particle up to a higher energy level, where it sees a much lower potential barrier and hence has a much higher tunneling rate $\Gamma_2 \gg \Gamma_1$.

Experimental proof of this quantization was already found in 1987 [33]. The energy distance of the lowest energy levels is given by the plasma frequency $\Delta E = \hbar\omega_{\mathrm{p}}$ and can be probed by microwave (MW) spectroscopy. If the MW frequency ω_{MW} is in resonance with the level spacing $\Delta E/\hbar$, the phase particle will be lifted to a higher state, where the potential barrier is lower and hence the tunneling rate is much higher (see Figure 2.3). In this case, a switching event will be recorded at a much lower bias current than expected in the absence of microwaves. With such a spectroscopy experiment, it is possible to experimentally observe the dependence of the plasma frequency on the bias current as given in (2.19).

For a deep potential well, its lower part can be approximated by a quadratic potential, which is equivalent to that of a harmonic oscillator. Consequently, the lower energy levels will have equal interlevel distances $\Delta E = \hbar\omega_{\mathrm{p}}$. This means that multi-photon transitions, where the condition

$$\omega_{\mathrm{MW}} = \frac{p}{q}\omega_{\mathrm{p}}\,, \quad p,q \in \mathbb{N} \tag{2.25}$$

is satisfied, will also be possible. This has been observed experimentally on a Josephson junction [39]. In the case of a shallow potential well, however, the anharmonicity becomes large and multi-photon transitions will less likely be observed.

2.3 The DC SQUID

The DC SQUID (**D**irect **C**urrent **S**uperconducting **Qu**antum **I**nterference **D**evice) takes advantage of the quantum mechanical nature of superconductivity, since it is based on the phase-sensitive summation of two supercurrents. A typical DC SQUID consists of a superconducting loop, which is interrupted by two identical Josephson junctions, each having a critical current I_{c} (see Figure 2.4a). Consequently, the bias current I can be described as

$$I = I_{\mathrm{c}}\sin\varphi_1 + I_{\mathrm{c}}\sin\varphi_2\,, \tag{2.26}$$

where $\varphi_{1,2}$ are the gauge-invariant phase differences across the junctions. Using a simple trigonometric identity, (2.26) can be transformed into

$$I = 2I_{\mathrm{c}}\sin\frac{\varphi_1 + \varphi_2}{2}\cdot\cos\frac{\varphi_1 - \varphi_2}{2}\,. \tag{2.27}$$

The wavefunction ψ going around the superconducting ring must be continuous, analogous to the discussion of the quantization of magnetic flux in Appendix A. Consequently, we obtain

$$\varphi_1 - \varphi_2 = 2\pi n + 2\pi\frac{\Phi}{\Phi_0}\,, \tag{2.28}$$

where Φ is the total flux penetrating the loop. A more detailed calculation can be found in [24, 26]. Inserting (2.28) into (2.27) yields an expression for the supercurrent through the SQUID as a function of the total magnetic flux traversing the loop:

$$I = 2I_{\mathrm{c}}\cos\left(\pi\frac{\Phi}{\Phi_0}\right)\cdot\sin\left(\varphi_1 + \pi\frac{\Phi}{\Phi_0}\right)\,. \tag{2.29}$$

We see that the critical current of the SQUID modulates with the cosine of the total flux through it and goes down to zero for $\Phi = (2m+1)\Phi_0/2$ (with m being an integer).

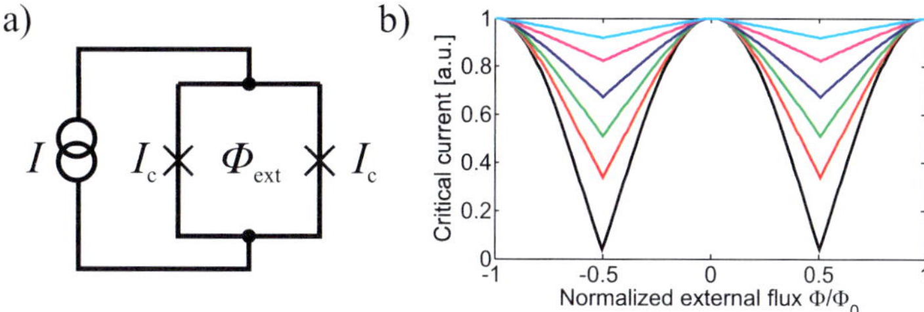

Figure 2.4: a) Schematics of a DC SQUID with two identical junctions. b) Numerically calculated modulation curves for different screening parameters β_L (from top to bottom: $\beta_L = 10$, 4, 1.8, 0.95, 0.5, 0.05).

However, due to the self-inductance L_{sq} of the SQUID loop, Φ is not equal to the externally applied flux Φ_{ext}. Instead, some flux is added by the circulating current I_{circ} induced in the ring, so that

$$\Phi = \Phi_{ext} - L_{sq}I_{circ} \, . \tag{2.30}$$

Since the current through the junctions is given by $I_c \sin \varphi_1 = I/2 + I_{circ}$ and $I_c \sin \varphi_2 = I/2 - I_{circ}$, respectively, (2.30) can be turned into

$$\Phi = \Phi_{ext} - \frac{L_{sq}I_c}{2} \left(\sin \varphi_1 - \sin \varphi_2 \right) . \tag{2.31}$$

Along the same line as for the current through the SQUID, this can be expressed as

$$\Phi = \Phi_{ext} - L_{sq}I_c \sin \left(\pi \frac{\Phi}{\Phi_0} \right) \cdot \cos \left(\varphi_1 + \pi \frac{\Phi}{\Phi_0} \right) . \tag{2.32}$$

This shows that depending on L_{sq} and I_c, a certain part of the external flux will be screened from the loop, so that the modulation depth of the SQUID's critical current will decrease with growing $L_{sq}I_c$ products. A dimensionless measure for this is the screening parameter

$$\beta_L = \frac{2L_{sq}I_c}{\Phi_0} \, . \tag{2.33}$$

In order to obtain the SQUID's critical current as a function of the external flux Φ_{ext}, (2.29) has to be maximized under the condition (2.32), which is not possible analytically. Numerically calculated SQUID modulation curves for various β_L values can be found in Figure 2.4b. More details about the design and fabrication of DC SQUIDs with photolithography as well as the corresponding measurements can be found in the *Studienarbeit* of Christian Goebel [Goe10].

2.4 The Superconducting Phase Qubit

The phase qubit is the simplest superconducting qubit, as it only consists of one Josephson junction. This is biased right below its critical current I_c, so that a shallow potential well is formed, and the two lowest energy levels are used as the qubit ground state $|0\rangle$ and the

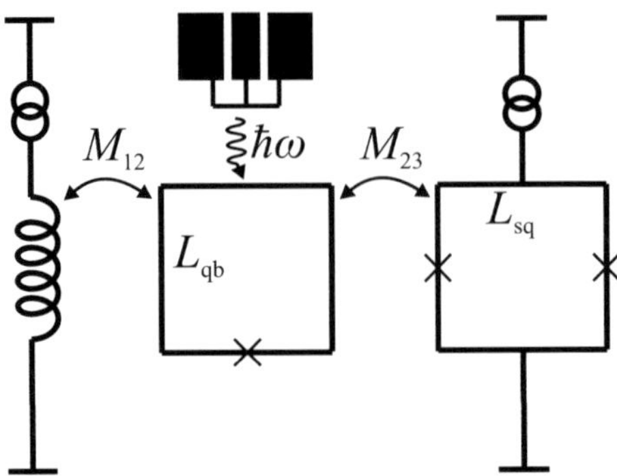

Figure 2.5: Schematics of a typical phase qubit, coupled to a bias line (on the left) with a mutual inductance M_{12}, and coupled to a read out DC SQUID (on the right) with a mutual inductance M_{23}. Qubit manipulations are carried out by microwave pulses.

qubit excited state $|1\rangle$. Qubit operations can be carried out with a resonant microwave pulse, while the sufficiently large anharmonicity in the shallow well ensures that the junction is not excited to higher energy levels. In order to read out the qubit, the washboard potential is tilted a little further, so that the tunneling rate from the lowest energy level is still negligible while the rate from the second-lowest level is relatively high (see Figure 2.3). In this way, the qubit state $|0\rangle$ is mapped to no switching and $|1\rangle$ is mapped to switching to the voltage state. Consequently, the occupation of the qubit states can be reconstructed with high fidelity by measuring the escape probability P_{esc}.

It is indeed possible to carry out phase qubit experiments on a directly biased Josephson junction [14, 22, 40, 41], but a rather complicated design of the bias lines is required to decouple the qubit from the environmental impedance of the bias lines Z_0 (for details see section 7.2.2). Furthermore, the junction switches to the voltage state when read out, so that heating occurs and quasiparticles are created. These problems can be solved with an elegant approach [42], in which the junction is placed in a superconducting ring with inductance L_{qb}. In this way, the qubit always remains in the superconducting state, while it is biased by an inductively coupled bias line and read out by an inductively coupled DC SQUID (see Figure 2.5). Furthermore, the impedance of the bias lines Z_0 is transformed into an effective impedance

$$Z_{\mathrm{eff}} = \left(\frac{L_{qb}}{M_{12}}\right)^2 Z_0\,, \qquad (2.34)$$

so that the junction is practically decoupled from the low-impedance environment. For the remainder of this thesis, the term *phase qubit* will refer to this kind of architecture.

Analogous to the DC SQUID described in section 2.3, the external flux Φ_{ext} through the phase qubit is partially screened. In this case, the screening parameter is given by

$$\beta_{L,qb} = \frac{2\pi L_{qb} I_c}{\Phi_0}\,. \qquad (2.35)$$

Since the junction in the phase qubit is not directly biased, but instead integrated into a superconducting ring, the last term in the washboard potential (2.17) has to be modified.

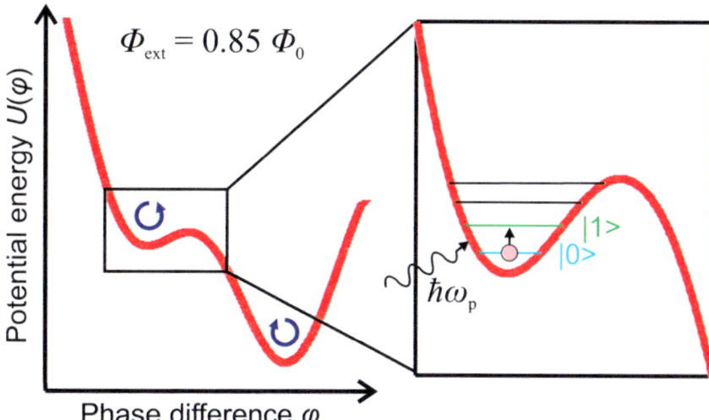

Figure 2.6: Schematic potential of a phase qubit at a typical working point. The two local minima correspond to circulating currents in opposite direction. The inset visualized how the qubit state is manipulated by resonant microwave radiation.

Now, the magnetic energy stored in the loop inductance L_{qb} has to be taken into account, which leads to [21]:

$$U\left(\varphi\right) = E_{\text{J}} \left(1 - \cos\varphi + \frac{(\varphi - 2\pi\Phi_{\text{ext}}/\Phi_0)^2}{2\beta_{L,\text{qb}}} \right) \tag{2.36}$$

This potential is shown in Figure 2.6 for a typical working point at $\Phi_{\text{ext}} = 0.85\Phi_0$. Qubit operation takes place in the shallow potential well on the left by resonant MW irradiation. The read out is performed analogously to the directly biased junction described above: A short flux pulse tilts the potential a little further, so that the excited state tunnels to the lower potential well. Since the two local minima correspond to opposite circulating currents, the qubit states are mapped to different flux signals, which are detected by the DC SQUID inductively coupled to the qubit loop (see Figure 2.5).

More details about the calculation of the phase qubit's energy levels, the theory of typical qubit experiments as well as the qubit manipulation and read-out procedure will be discussed in chapter 9.

3 Quality Criteria for Short Nb/Al-AlO$_x$/Nb Josephson Junctions

High-quality Josephson junctions are generally desirable for all applications, but especially crucial for the quantum experiments performed in this thesis. This chapter explains how short Nb/Al-AlO$_x$/Nb junctions (i.e. having lateral dimensions smaller than λ_J, see (2.13)) are characterized and how the typical quality parameters are defined. It also aims at revealing the physics behind these characteristic measurements in order to find out how the fabrication process has to be modified for an improved junction quality. The chapter starts by discussing typical current-voltage characteristics and continues with detailed considerations about the individual quality parameters. Finally, the dependence of the critical current on an external magnetic field and its meaning for junction characterization are addressed.

3.1 Current-Voltage Characteristics

For a Josephson junction, the amount of information about its quality extractable from its current-voltage characteristics strongly depends on the value of the Stewart-McCumber parameter (2.21). For overdamped junctions having a Stewart-McCumber parameter $\beta_C \ll 1$, not much information can be obtained. For underdamped junctions with $\beta_C \gg 1$ in contrast, a lot of information about their quality can be extracted from the IV curves. This shall be explained in detail in the following.

Figure 3.1a shows the current-voltage characteristics of a junction with a Stewart-McCumber parameter of $\beta_C = 0.1$, which was reached by shunting the junction with a parallel resistance $R_{sh} = 0.4\,\Omega$, being much smaller than the junction resistance R. The influence of damping on the dynamics of this junction can be illustrated using the washboard potential of Figure 2.2: Once the critical current I_c is exceeded, the quasiparticles which contribute to charge transport only see a small resistance R_{sh}, so that the IV curve converges quickly towards the ohmic curve $V = R_{sh} \cdot I$. This is equivalent to the phase particle running down the washboard potential (2.17) with a velocity $\dot{\phi} \propto V$ which is proportional to the slope I. When the current is decreased, the particle will stop running as soon as a shallow potential barrier is formed for $I \lesssim I_c$, since its kinetic energy is quickly absorbed by the high damping. Consequently, no hysteresis in the current-voltage characteristics is observed. This means that only the critical current and the shunt resistance can be extracted, as can be seen in Figure 3.1a. These two parameters do not give any information about the inherent quality of the junction.

In this thesis, the undisturbed quantum dynamics in Josephson junctions and in circuits containing them shall be investigated, which means that highly underdamped samples with $\beta_C \gg 1$ are required. Additionally, this is also the case in which junction quality parameters can be evaluated from the IV curves. In contrast to an overdamped junction, the

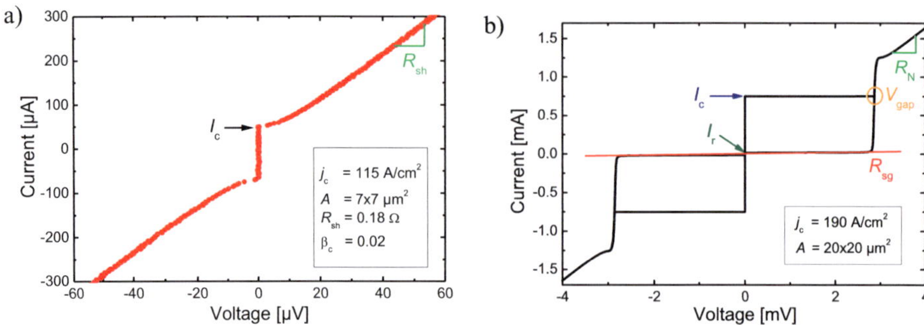

Figure 3.1: a) Typical *IV* curve of an overdamped junction. No quality parameters can be evaluated. b) Typical *IV* curve of an underdamped junction. Here, the quality parameters can be extracted, as described in the text.

current-voltage characteristics of an underdamped structure show a strong hysteresis (see Figure 3.1b). This is due to the fact that the kinetic energy which the phase particle acquires in the running state is not damped out after the current is lowered past I_c, so that its momentum can carry the particle over the rising potential barrier and the movement only stops for a flat potential at $I_r \approx 0$. This is commonly known as retrapping to the zero-voltage state and I_r is called the retrapping current.

The occurrence of this hysteresis allows the observation that the quasiparticle resistance R defined in the RCSJ model (2.14) does actually depend on the voltage. Typically, two resistances in different regimes are defined: For voltages below the superconducting energy gap $V < 2\Delta/e$, quasiparticle conduction is only possible through thermal excitation, multi-photon processes or pinholes in the tunneling oxide, so that the rather high subgap resistance R_{sg} is seen. For voltages $V > 2\Delta/e$, however, the energy supplied by the current source is sufficient to break up Cooper pairs, so that quasiparticles from one side can tunnel to unoccupied states on the other side of the barrier. In this ohmic regime, the normal resistance R_N is observed. It is now clear that the Stewart-McCumber parameter β_C can be used as a measure for the strength of the hysteresis in the *IV* curve if the resistance R in (2.21) is replaced by R_{sg}.

It can be seen in Figure 3.1b that a large voltage jump occurs at the critical current I_c. Since at the gap voltage $V_{gap} = 2\Delta/e$, sufficient energy for breaking of Cooper pairs is supplied, the voltage jump stops here and does not continue to the voltage drop for ohmic quasiparticle transport in the subgap regime $V = R_{sg} \cdot I_c$, which would be much higher.

3.1.1 The Gap Voltage

As the gap voltage is directly related to the energy gap $2\Delta(T)$, it reflects the superconducting properties of the Nb electrodes. The BCS theory predicts that $2\Delta(T = 0)$ can be calculated using $2\Delta = 3.528\,k_B T_c$ [24], where k_B is the Boltzmann constant. This is commonly known as the weak-coupling limit. Niobium, however, belongs to a class of superconductors in which the electron-phonon interaction is particularly strong, which is commonly known as the strong-coupling limit. Here, the BCS formula does not hold, so that the ratio $2\Delta/(k_B T_c)$

has to be determined individually for each material. For Nb, the literature values for $2\Delta(0)$ are between 3.06 and 3.14 meV [43] and the critical temperature accounts for $T_\mathrm{c} = 9.25\,\mathrm{K}$, so that $2\Delta/(k_\mathrm{B}T_\mathrm{c})$ lies between 3.84 and 3.94.

The temperature dependence of the superconducting energy gap can only be calculated numerically. However, two approximations for different temperature regimes were found. For $T < 0.5T_\mathrm{c}$, it is given by [44]

$$\frac{\Delta(T)}{\Delta(0)} \approx 1 - 3.33 \left(\frac{T}{T_\mathrm{c}}\right)^{1/2} \exp\left(-1.76 \cdot \frac{T_\mathrm{c}}{T}\right), \tag{3.1}$$

while for $T \approx T_\mathrm{c}$, the temperature dependence can be approximated as [24]

$$\frac{\Delta(T)}{\Delta(0)} \approx 1.74 \left(1 - \frac{T}{T_\mathrm{c}}\right)^{1/2}. \tag{3.2}$$

Strictly speaking, these formulas are only valid in the weak-coupling regime, but they are a good approximation in most cases [24]. As can be seen in Figure 3.2, the temperature range $0 < T < 4.2$ K investigated in this thesis can be described by (3.1) in good approximation. Employing (3.1) and the literature values of $2\Delta(0)$ for Nb, we can estimate the theoretical value for the gap voltage at 4.2 Kelvin to be in the range of

$$V_\mathrm{gap,theo} = 2.91\,\mathrm{mV} \ldots 2.99\,\mathrm{mV}. \tag{3.3}$$

If the superconducting electrodes are made of thin films having a thickness smaller than the London penetration depth λ_L, their superconducting properties will be degraded compared to those of ideal bulk niobium. In particular, the energy gap and hence also the critical temperature will be significantly decreased. Values for λ_L at zero temperature can be found in literature [45, 46] and account for $\lambda_\mathrm{L}(0) = 85$ nm. Considering the temperature dependence $\lambda_\mathrm{L}(T) = \lambda_\mathrm{L}(0)/\sqrt{1 - (T/T_\mathrm{c})^4}$ [24], this gives a value of $\lambda_\mathrm{L} = 87$ nm at 4.2 Kelvin,

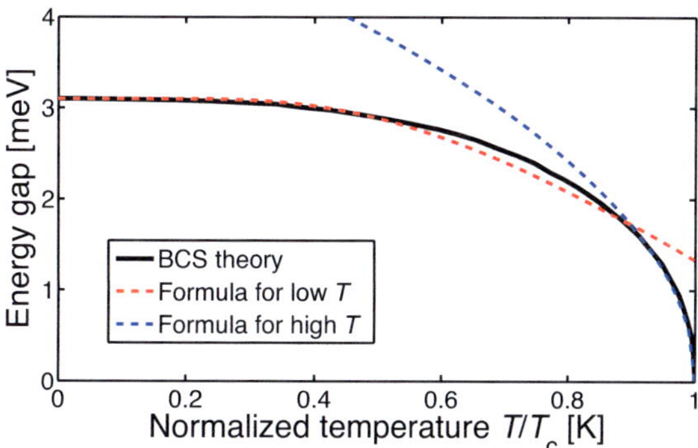

Figure 3.2: Temperature dependence of the superconducting energy gap for niobium, with the mean literature value $2\Delta(0) = 3.1$ meV. The black curve was extracted from [24], where $2\Delta(T)$ was calculated numerically in the BCS weak-coupling limit. The red curve corresponds to (3.1) and the blue curve to (3.2).

which is in good agreement with values of $\lambda_L = 86$ nm found experimentally at the same temperature [47].

For the fabrication technology, this means that the superconducting electrodes should have a thickness larger than ≈ 90 nm. Furthermore, the film deposition should be optimized towards minimal film stress, which should lead to good superconducting properties [48] and hence critical temperatures close to the theoretical value of $T_c = 9.25$ K. Additionally, in [49] it was found that the gap voltage was strongly affected by the number of impurities included in the Nb films, which were caused by the residual gas in the vacuum chamber before the trilayer deposition. Consequently, a high vacuum in the trilayer deposition system as well as a clean sputtering gas are important.

Experimentally, values of $V_{gap} = 2.95$ mV have been reached at $T = 4.2$ K in studies directly focusing on maximizing the gap voltage [49], but generally values of $V_{gap} > 2.8$ mV at this temperature are already considered to indicate a very high junction quality.

3.1.2 The $I_c R_N$ Product

Since the critical current I_c of a Josephson junction scales with the junction's dimensions just as the inverse of its resistance in the normal state $1/R_N$, the product $I_c R_N$ should have an invariant value. It can be shown within the Ginzburg-Landau theory that this value generally depends only on material and temperature, but not on the junction geometry. This is true even for constriction type junctions, as long as their linear dimensions are smaller than the coherence length $\xi(T)$ [24].

For tunnel junctions as the Nb/Al-AlO$_x$/Nb devices discussed in this thesis, Ambegaokar and Baratoff used the BCS theory to derive an exact result for the full temperature dependence [50]:

$$I_c R_N = \frac{\pi \cdot 2\Delta(T)}{4e} \tanh\left[\frac{2\Delta(T)}{4k_B T}\right] \tag{3.4}$$

This temperature dependence was calculated using the black curve from Figure 3.2 for $2\Delta(T)$ and is shown in Figure 3.3. Using the mean of the literature values for the gap, $2\Delta(0) = 3.1$ meV, as well as equations (3.1) and (3.4), the theoretical value at 4.2 Kelvin

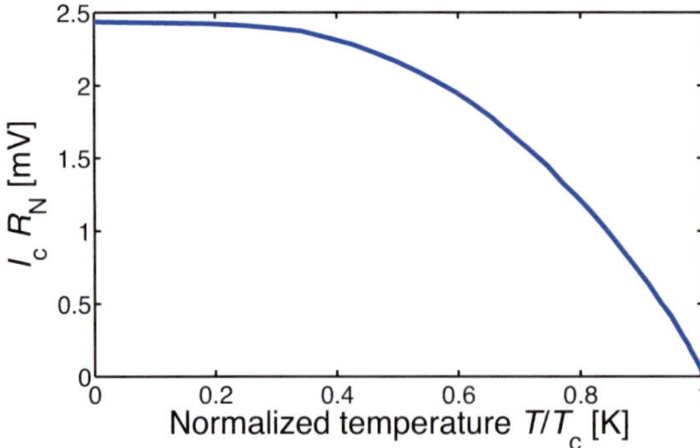

Figure 3.3: The temperature dependence of the critical current after Ambegaokar and Baratoff [50] for a Josephson junction with Nb electrodes.

can be calculated as $I_cR_N = 2.24$ mV. However, for typically observed gap voltages of $V_{gap} = 2.8$ mV, the theoretically expected value accounts for $I_cR_N = 2.11$ mV.

Since this expression describes ideal Cooper pair tunneling, it is never reached in real samples. Reasons for this are that the tunneling layer is not a rectangular potential barrier, but rather varies in all three spatial dimensions and has sloped potential walls. Nevertheless, the measured I_cR_N product is an important parameter to determine the strength of Cooper pair tunneling in a real Josephson junction. Experimentally, values above $I_cR_N = 1.5$ already indicate a good tunneling barrier.

However, a high I_cR_N product can only be used as a quality parameter if no excess current is observed. The latter is present if the straight line given by the differential normal resistance in the first quadrant of the IV curve intersects the current axis at a positive value I_{exc}. In this case, quasiparticle conduction contributes to the critical current, so that I_cR_N is not a measure for the strength of Cooper pair tunneling anymore.

Technologically, to come close the ideal tunneling described by (3.4), the AlO_x barrier should be well-defined, meaning that it is homogeneous in space and does not have impurities. Additionally, high temperatures over $200\,°C$ during processing should be avoided, as they lead to lowered critical currents [51].

3.1.3 The Subgap Branch

In contrast to the normal resistance R_N, which is voltage-independent, the subgap resistance $R_{sg}(V)$ is a non-trivial function of the voltage V. Although it is clear that a low leakage current, i.e. a high subgap resistance is desirable for all applications, the complex behavior of $R_{sg}(V)$ makes it quite unclear at which voltage R_{sg} values should be evaluated for a quality parameter. In the following, different physical processes which contribute to quasiparticle conductance below 2Δ are discussed and related to certain applications and the corresponding quality parameters. Finally, the technological requirements for high subgap resistance values will be discussed.

Classic Quality Parameters

For many applications such as RSFQ, SQUIDs or voltage standards, the typical operation temperature is 4.2 Kelvin. In the Gorter-Casimir two-fluid model [52, 53], the temperature dependence of the Cooper pair density n_s is given by

$$n_s = n_s(0)\left[1 - \left(\frac{T}{T_c}\right)^4\right]. \tag{3.5}$$

This means that in Nb at $T = 4.2$ K, more than four percent of the Cooper pairs are already broken up into quasiparticles. Consequently, a certain subgap conductance cannot be avoided even for junctions of very high quality, and fine structures of the subgap regime like Multiple Andreev Reflection (discussed below) cannot be observed at this temperature.

Since the devices mentioned above are all operated in the voltage regime, low leakage currents also at higher voltages below the gap are important. Hence, the subgap resistance is conventionally evaluated at a voltage of 2 mV. This is also advantageous regarding the required measurement technique, since IV characteristics are usually recorded with a current bias and very few points are obtained on the retrapping branch for high R_{sg} values.

The subgap resistance determined in this way is commonly used for two quality parameters. One is the ratio of subgap to normal resistance $R_{\mathrm{sg}}/R_{\mathrm{N}}$ and the other one is the characteristic voltage $V_{\mathrm{m}} = I_{\mathrm{c}}R_{\mathrm{sg}}$. Both values are independent of the junction geometry so that they allow comparison between all junctions made from the same material. For the applications mentioned earlier, values of $R_{\mathrm{sg}}/R_{\mathrm{N}} > 10$ at 4.2 Kelvin are usually considered to indicate a high quality. The characteristic voltage V_{m} combines the information of the $I_{\mathrm{c}}R_{\mathrm{N}}$ product and the $R_{\mathrm{sg}}/R_{\mathrm{N}}$ ratio, so that it is often used as the single quality parameter for a trilayer process. Here, values of $V_{\mathrm{m}} > 30$ mV are typical for a high junction quality.

Multiple Andreev Reflection

According to (3.5), quasiparticles are almost entirely frozen out below $T_{\mathrm{c}}/10$, which corresponds to $T \approx 0.9$ K for niobium. Consequently, contributions to the subgap conductance which are not due to thermally excited quasiparticles can be observed below this temperature. An important phenomenon, which was first observed experimentally in the early 1960s [54, 55], is the sub-harmonic gap structure (SGS). It consists of current steps (i.e. peaks in $\mathrm{d}I/\mathrm{d}V$) at voltages $2\Delta/(ne)$, where n is an integer. A typical measurement curve can be seen in Figure 3.4. One suggested explanation is the one of Josephson self-coupling (JSC) [56], which relies on the nonlinear coupling of the Josephson radiation (produced by the AC Josephson current) with this oscillating Josephson current itself, creating a DC component of current below 2Δ. In this model, the current steps should be limited to odd n values and the step height should depend on magnetic field [57]. Especially the former is not observed experimentally, so that JSC is usually ruled out as an explanation for typical SGS.

A competing explanation is given by multi particle tunneling (MPT) [58] or multiple Andreev reflection (MAR) [59], which are physically equivalent [60]. However, MAR gives a

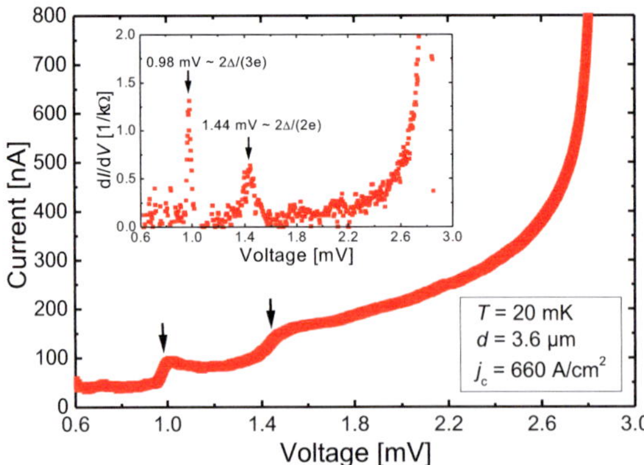

Figure 3.4: The measurement of an *IV* curve with a voltage bias setup allows the observation of MAR. The inset shows that the positions of the $\mathrm{d}I/\mathrm{d}V$ peaks correspond well to the observed gap voltage $V_{\mathrm{gap}} = 2.91$ mV. The measurement was carried out on a circular junction with a diameter of 3.6 µm at the Institut für Festkörperphysik (IFP) within the frame of this thesis.

more general and more physical picture [60], so that MAR has prevailed as the commonly used explanation today. It involves the process of Andreev reflection [61], which means that an electron in the insulator having an energy smaller than the gap Δ can create a Cooper pair in the superconductor while a hole is reflected. So far, no effective charge transport has occurred. MAR suggests that such a reflection can happen multiple times at the opposite insulator/superconductor interfaces, so that a quasiparticle, for example coming form the left electrode, is reflected back and forth between the right and the left superconductor/insulator interfaces. Once its energy is sufficient to reach the unoccupied quasiparticle states in the right superconducting electrode, an effective charge transport has occurred. It can be seen that for $n-1$ reflections, the process will be possible at a voltage $2\Delta/(ne)$, while its probability will decrease with rising n since the electron has to pass the tunneling barrier n times. This is why MAR steps are usually only observed experimentally for low values of n.

Retrapping to the Zero-Voltage State

Measuring the *IV* characteristics of a Josephson junction with a voltage bias results in a much a higher resolution of the subgap regime. For zero magnetic field, the current rises back up for lowering voltage before the junction switches back to the voltage state (see Figure 3.5). It was also found that for lowering temperature, the retrapping current only decreases down to $T = 1$ K and saturates at this value [62]. Both effects are equivalent and have the consequence that the intrinsic subgap resistance (i.e. the resistance in the zero-voltage state) cannot be measured without applying a magnetic field.

The behavior of the JJ close to retrapping and also the value of the retrapping current can be described by using a simple energy-balance argument [63]. In this model, the power input *IV* from the DC current source is dissipated by currents at two frequencies: The first term, V^2/R_{QP} is dissipated within the junction by the DC quasiparticle current while

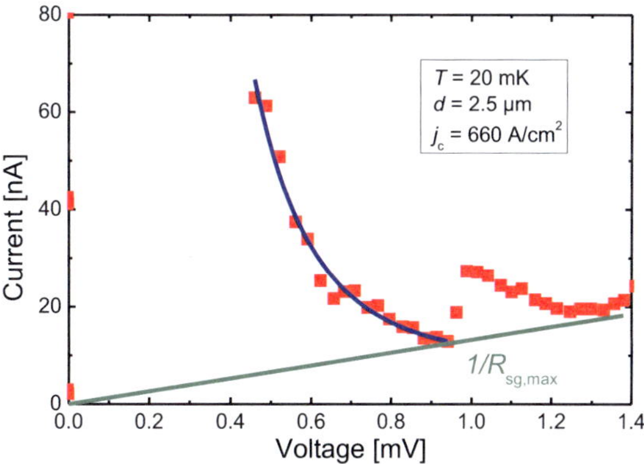

Figure 3.5: Retrapping to the zero-voltage state in a voltage bias setup for a circular junction with a diameter of 2.5 µm. The blue curve is obtained by a fit with equation (3.8) while the inverse of the slope of the green line is the maximal subgap resistance $R_{sg,max}$. The measurement was carried out at the Institut für Festkörperphysik (IFP) within the frame of this thesis.

R_{QP} is the resistance these quasiparticles see. The second term is given by the dissipation at the Josephson frequency (2.11), meaning that an ac supercurrent $I_{c0}\sin\omega_J t$ is flowing through a complex impedance $Z(\omega_J)$. Consequently the dissipated power at ω_J is given by $(I_{c0}{}^2/2)\mathrm{Re}(Z(\omega_J))$. Both contributions combined lead to the expression for the current [24]

$$I(V) = V/R_{QP} + (I_{c0}{}^2/2V)\mathrm{Re}(Z(\omega_J)).\tag{3.6}$$

The impedance $Z(\omega_J)$ is obtained from the parallel contributions of the junction capacitance C, the quasi-particle resistance R_{QP} and the impedance of the lines leading to the junction Z_0. The latter is considered to be real and in the range of $50 - 100\ \Omega$ [33]. Hence we get

$$\frac{1}{Z(\omega_J)} = i\omega_J C + \frac{1}{R_{QP}} + \frac{1}{Z_0}.\tag{3.7}$$

If we take into account that $Z_0 \ll R_{QP}$, calculate the real part of $Z(\omega_J)$, remember that $\omega_J = 2eV/\hbar$ and put all this into (3.6), we obtain the expected $I(V)$ behavior in the subgap branch close to retrapping [64]:

$$I(V) = \frac{V}{R_{QP}} + \frac{I_{c0}{}^2}{2V}\cdot\left[\frac{Z_0}{1+(2e/\hbar\cdot CZ_0 V)^2}\right].\tag{3.8}$$

The validity of (3.8) for voltages below the highest-order MAR step has been experimentally confirmed [63, 64] and can be seen by the excellent agreement between data and fit curve in Figure 3.5.

Subgap Leakage and Coherence Times

The quasiparticle resistance R_{QP} in (3.8) describes the power dissipated at the Josephson frequency in the voltage state. For quantum devices, which are exclusively operated in the zero-voltage state, however, the damping within the potential well at the plasma frequency is relevant. Consequently, it is desirable to determine an intrinsic subgap resistance $R_{sg,intr}$.

In order to get a measure for the damping in the zero-voltage state, the straight line of minimal slope which still touches the subgap branch of the *IV* curve (as illustrated by the green line in Figure 3.5) is often evaluated. The inverse of this slope is the maximal subgap resistance $R_{sg,max}$ [64,65]. However, due to the rising current for decreasing voltage discussed in the previous section and the fact that retrapping occurs already at voltages $V \approx 0.5$ mV (see Figure 3.5), the $R_{sg,max}$ values observed at zero magnetic field are not equal to the actual intrinsic subgap resistance $R_{sg,intr}$.

Only by entirely suppressing the critical current with a magnetic field H, the AC supercurrent and hence the second term in (3.8) will disappear, so that the remaining linear term $V = R \cdot I$ should give a measure for $R_{sg,intr}$ [64, 65]. In other words, for an entirely suppressed critical current, it will not be possible for the curve in Figure 3.5 to rise to higher current values and jump back to the superconducting branch at $I > 0$. It has been shown that even for critical currents suppressed as far as $I_c(H) = 10^{-4}I_c(0)$, only a slight saturation in $R_{sg,max}$ is observed [64] (see Figure 5.14b). However, the values of $R_{sg,max}$ at such low critical currents can be used as a lower boundary for $R_{\text{sg,intr}}$ [64,65]. This value allows to estimate the decoherence rate due to subgap conductance in superconducting quantum bits.

If subgap leakage was the only mechanism responsible for decoherence in a superconducting qubit, the relaxation time T_1 would be given as the ring-down time [64–66]

$$T_1 = R_{\text{sg,intr}}C, \tag{3.9}$$

while the dephasing time T_2 induced by thermal noise would be calculated as [64–66]

$$T_2 = \frac{2\hbar^2 R_{\text{sg,intr}}}{\Phi_0^2 k_B T}. \tag{3.10}$$

Here, C is the junction capacitance, $\hbar$ the reduced Planck constant, Φ_0 the magnetic flux quantum and k_B the Boltzmann constant. A detailed explanation of the physical significance of T_1 and T_2 is given in section 9.3. It can be seen directly that a high intrinsic subgap resistance is crucial for the general operability of superconducting qubits.

Technological Requirements for Low Subgap Leakage

Since low subgap leakage currents are important or even crucial for most Josephson junction applications, special care must be taken to fulfill this requirement. Since stress induced damage to the tunneling barrier has been found to decrease the subgap resistance [51], the stress in the Nb electrodes of the SIS trilayer has to be minimized. Furthermore, heating the junctions to over 200 °C leads to a higher leakage [51], so that high temperatures during processing should be prevented. Additionally, any defects or pinholes in the tunneling barrier or the insulating layers should be avoided, since they would allow quasiparticle transport from the top to the bottom electrode. For the dielectric layers, achieving good insulating properties is especially difficult at the edges of the lower Nb electrode, so that in design of the upper Nb electrode, the length of the overlap above these edges should be minimized.

3.2 $I_c(\Phi)$ Modulation

A schematic picture of a short Josephson junction (having dimensions much smaller than the London penetration depth λ_J, see (2.13)) in a magnetic field is shown in Figure 3.6.

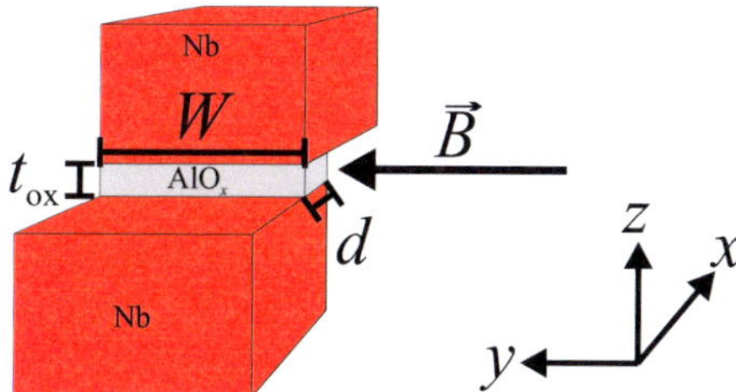

Figure 3.6: Schematics of a Josephson junction with area $w \cdot d$ in a magnetic field B along the coordinate y.

The junction area $w \cdot d$ is in the (x,y)-plane while the magnetic field $\vec{B}$ is oriented along the y-axis. The magnetic field can penetrate the insulating barrier of thickness t_{ox} and the adjacent superconducting electrodes down to a distance λ_{L} from the barrier. Consequently, the magnetic flux penetrating the junction is given by $\Phi = Bd(t_{\text{ox}} + 2\lambda_{\text{L}})$.

Generally, the critical current density $j_{\text{c}}(x,y)$ is a function of space. Since $\vec{B}$ is directed along the y-axis, only changes of $j_{\text{c}}(x,y)$ in x-direction will influence the magnetic field dependence of the critical current. Hence, the integral critical current density

$$\Upsilon_c(x) = \int\limits_{-w/2}^{w/2} j_{\text{c}}(x,y)\mathrm{d}y \tag{3.11}$$

is defined. It can be shown (a complete derivation can be found in [24–26]) that the critical current of the junction as a function of the magnetic flux is given by

$$I_{\text{c}}(\Phi) = \left| \int\limits_{-\infty}^{\infty} \Upsilon_c(x)\exp\left(i\frac{2\pi\Phi}{\Phi_0}\frac{x}{d}\right)\mathrm{d}x \right|. \tag{3.12}$$

In other words, the $I_{\text{c}}(\Phi)$ modulation is given as the absolute value of the Fourier transform of the integral critical current density $\Upsilon_c(x)$. For an ideal rectangle-shaped Josephson junction, the Fourier transform can be calculated analytically and leads to

$$I_{\text{c}}(\Phi) = I_{\text{c}} \left| \frac{\sin\left(\pi\frac{\Phi}{\Phi_0}\right)}{\pi\frac{\Phi}{\Phi_0}} \right|. \tag{3.13}$$

This modulation curve can be seen in Figure 3.7a. It is commonly known as Fraunhofer pattern, since it is equivalent to the diffraction pattern of light at a single slit. The corresponding critical current density distribution can be seen in the inset of the same figure.

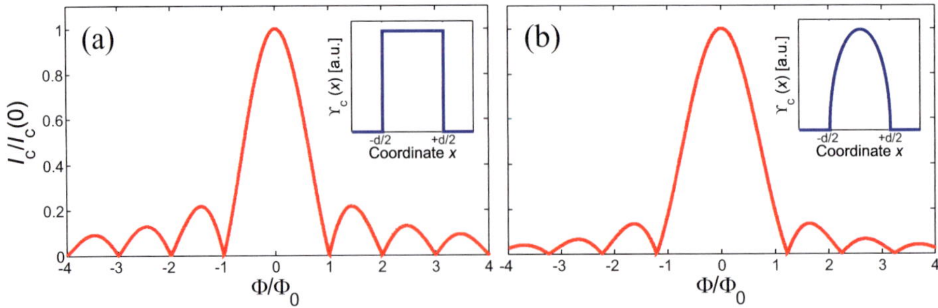

Figure 3.7: Simulation of $I_{\text{c}}(\Phi)$ modulations for (a) a square junction and (b) a circular junction. The red curves were created by Fourier transformation of the blue curves shown in the insets.

Also for an ideal circular Josephson junction with diameter d, an analytical solution of (3.12) is possible and

$$I_c(\Phi) = I_c \left| \frac{J_1\left(\pi \frac{\Phi}{\Phi_0}\right)}{\frac{1}{2}\pi \frac{\Phi}{\Phi_0}} \right| \tag{3.14}$$

is obtained as the critical current modulation [25]. Here, $J_1(x)$ is the Bessel function of the first kind and the magnetic flux is again given by $\Phi = Bd(t_{ox} + 2\lambda_L)$. The critical current density distribution and the corresponding modulation curve of a circular junction (also called Airy pattern) can be seen in Figure 3.7b. In contrast to a rectangular junction, the side maxima are of much smaller amplitude and the minima are not at integer multiples of the flux quantum Φ_0.

$I_c(\Phi)$ measurements are an important tool for the characterization of junction quality. If the obtained $I_c(\Phi)$ modulation deviates from (3.13) or (3.14), respectively, the spatial critical current distribution $j_c(x,y)$ differs from the expected behavior. Such deviations might be due to an inaccurate definition of the junction geometry (i.e. patterning problems) or spatial inhomogeneities in the tunneling barrier.

4 Fabrication Technology

In this chapter, the development and optimization of the Nb/Al-AlO$_x$/Nb Josephson junction fabrication technology is being discussed. First, an overview of the fabrication process and the Josephson junction quality before the start of this thesis is given and potential process improvements are considered. Subsequently, the physical background discussed in chapter 3 is used to find the optimal process solutions for the fabrication of high quality junctions for quantum experiments. Furthermore, the processing steps which were optimized for the standard photolithography process are discussing in detail. Afterwards, these changes are summarized in an overview of the newly developed process. Finally, the requirements for a fabrication process for high quality sub-µm junctions are considered and the single optimized or newly developed processing steps are discussed in detail. Some parts of this chapter have been published in [KMI$^+$11].

4.1 Initial Technological Process at the IMS

A schematic overview of the Josephson junction fabrication process from the time before the start of this PhD thesis can be seen in Figure 4.1. At first, the Nb/Al-AlO$_x$/Nb trilayer is created in a DC magnetron sputtering system with layer thicknesses of 200 nm (lower Nb electrode), 7 nm (Al) and 100 nm (top Nb electrode). Oxidation of Al is carried out for 30 min at room temperature in a pure oxygen atmosphere. Then, the Josephson junctions are defined by positive photolithography and reactive-ion-etching (RIE). A first insulating layer is created using the same resist mask by anodic oxidation. Afterwards, the shape of

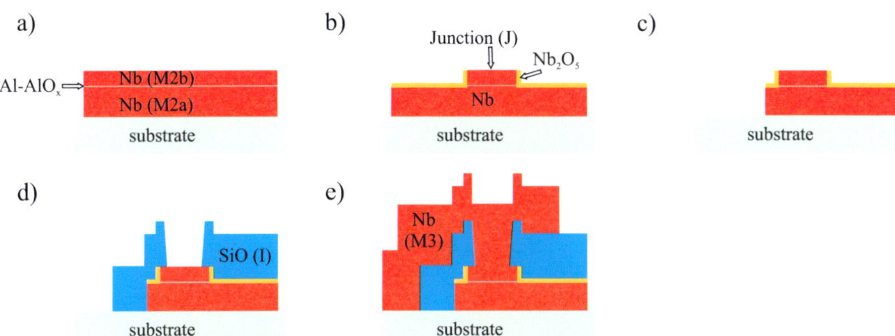

Figure 4.1: Schematic overview of the Josephson junction process from before the start of this thesis: a) As-deposited trilayer b) After junction definition and anodic oxidation c) After patterning of bottom electrode d) With deposited SiO insulation layer e) Final structure with wiring layer.

the bottom electrode is defined by positive photolithography and subsequent RIE, ion beam etching (IBE) and RIE. The intermediate IBE step is necessary because the AlO_x layer cannot be etched by RIE. In the following, negative photolithography and thermal evaporation of SiO are employed to create the second insulating layer. Finally, the wiring layer is defined by negative photolithography and DC magnetron sputtering of Nb. A detailed list of the processing parameters of this old process including the names of the layers can be found in Appendix B.1.

4.1.1 Junction Quality Before Start of Thesis and Potential for Process Improvement

In order to discover processing steps which had to be improved, the quality of the Josephson junction technology described in the previous section was analyzed. For this purpose, the measurement data of test chips from five wafers (which were fabricated in October 2005 and characterized at 4.2 Kelvin in December 2005) were evaluated concerning yield and quality parameters. Altogether, *IV* characteristics of 29 junctions of sizes 5×5 μm^2, 10×10 μm^2 and 20×20 μm^2 were analyzed. A typical *IV* curve of a 10×10 junction can be seen in Figure 4.2.

The results of the analysis are given in Table 4.1. First, it is obvious that the fabrication of junctions with sizes below 20×20 μm^2 had a poor yield. Furthermore, the junction quality seems to reduce with the junctions' dimensions. It is doubtful that a reduced quality for 10×10 junctions is already due to alignment problems, so that the size dependent quality might indeed reveal flaws in the process. This would be a significant problem for the quantum experiments performed within this thesis, since here, junction sizes around and below 3 μm are required. Generally, it is desirable to reliably fabricate Josephson junctions down to 5×5 μm^2 with photolithography, since the latter is a lot less time consuming than electron-beam lithography and such junctions can be used for many applications as well as junction characterization.

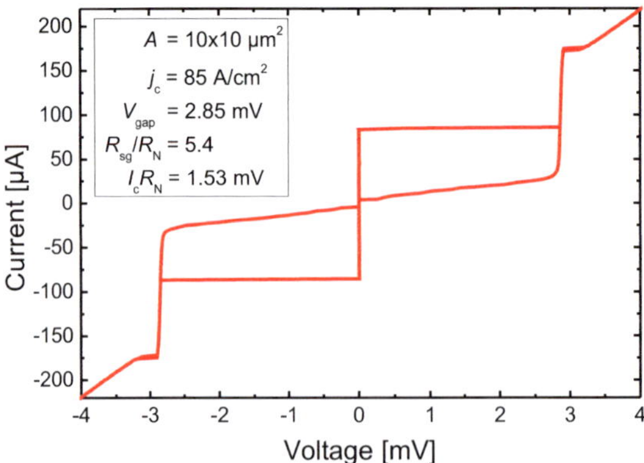

Figure 4.2: Typical *IV* curve of a Josephson junction with area $A = 10 \times 10$ μm^2, fabricated before the start of this thesis and measured at 4.2 Kelvin.

Table 4.1: Yield and quality of samples fabricated before the start of this thesis. $\langle\rangle$ denotes averages.

size [μm^2]	working	defective	$\langle V_{gap}\rangle$ [mV]	$\langle R_{sg}/R_N\rangle$	$\langle I_c R_N\rangle$	$\langle V_m\rangle$ [mV]
5×5	0	11	-	-	-	-
10×10	5	4	2.82	8.5	1.54	12.7
20×20	8	1	2.82	12.4	1.71	21.4

Concerning the junction quality, the rather high gap voltages $V_{gap} = 2.8$ mV indicate a high quality of the Nb electrodes. The R_{sg}/R_N ratios around 10 might be sufficient for many applications such as RSFQ, SQUIDs, etc., but not for quantum measurements and especially quantum bits, where the intrinsic damping has to be extremely low for long coherence times (see section 3.1.3). Hence, the improvement of the R_{sg}/R_N ratio will be one of the main points of this chapter. The $I_c R_N$ products of around 1.6 are reasonably high, but it is desirable to improve them for a stronger Cooper pair tunneling.

In order to find possible solutions for process improvement, the findings gained in junction characterization were related to the fabrication process discussed in the precedent section. In the following list, potentially problematic issues and possible solutions are given:

- Although any Nb film thickness $> \lambda_L$ should be sufficient for good superconducting properties (see section 3.1.1), the thickness of the lower electrode (M2a) accounted for 200 nm. This might lead to poor step coverage of the insulating SiO layer, which would cause a decreased yield and a higher subgap leakage.
- There might be stress in the Nb films, which would lead to a decreased subgap resistance [51]. Hence, the Nb film stress should be carefully minimized.
- Due to the affinity of Al to Nb, the Al layer planarizes the rather rough Nb surface [67, 68]. This effect was found to be optimal for an Al thickness of 7 nm [69]. Since an inhomogeneous tunneling oxide might lead to lowered $I_c R_N$ products, higher subgap leakage and a decreased reproducibility, the Al sputtering rate should be defined carefully to indeed obtain a thickness of 7 nm.
- As can be seen in Figure 4.1, the sidewalls of the lower electrode (M2a) are not anodically oxidized. This might result in spurious leakage currents and hence lead to a lowered subgap resistance and a lower yield. This problem can be solved by simply changing the order of the processing steps.
- The parameters used for RIE led to an etching rate of around 560 nm/min. This makes over-etching very likely to occur, which leads to a badly defined junction size and hence lower reproducibility. Furthermore, the RIE process left some resist residue on the chip, which was very hard to clean off. This means that the anodic oxidation might be incomplete in some spots, leading to a lowered subgap resistance. Furthermore, electric contact between M2b and M3 might be prevented for small junctions. Consequently, new etching parameters should be found, which result in cleaner etching with a lower rate.

4.2 Optimization of Standard Photolithography Process

In this section, the conclusions attained in the precedent analysis shall be used to improve the standard photolithography process, in order to get a high junction quality for samples down to sizes of 5 µm. Such junctions can be used for many applications, such as RSFQ [2], voltage standards [3, 4], digital-to-analogue converters [7] and SQUIDs [8]. Furthermore, such a technology can be used to perform quantum experiments on long junctions with one dimension larger than the Josephson penetration depth λ_J, defined in (2.13). One example for this is the investigation of the dynamics of fluxons or fractional fluxons, which was carried out in the *Studienarbeit* and *Diplomarbeit* of Max Meckbach within the frame of this thesis. Details about the physics behind these experiments and the special technological requirements as well as interesting measurement results can be found in [Mec08, Mec09].

Detailed parameters of all discussed processing steps can be found in Appendix B.2.

4.2.1 Trilayer Deposition

First, the thickness of the M2a Nb layer was reduced from 200 nm to 100 nm. This is sufficiently larger than the London penetration depth $\lambda_L = 87$ nm, which is required for high gap voltages (see section 3.1.1). The fact that a gap voltage of $V_{\text{gap}} = 2.95$ mV was reached for a Nb film thickness of 100 nm in [49] shows that such a Nb thickness can indeed lead to good results.

Working Point for Minimal Nb Film Stress

As discussed above, stress-free Nb films should lead to an increased subgap resistance [51] and to a high gap voltage [48]. Consequently, the working point for minimal film stress was determined.

As the main parameter influencing the stress is the Ar working pressure p_{Ar} during sputtering [51], the latter was varied systematically and the film stress was evaluated. First, this was done by simply sputtering Nb films onto photoresist, so that the films could relax on

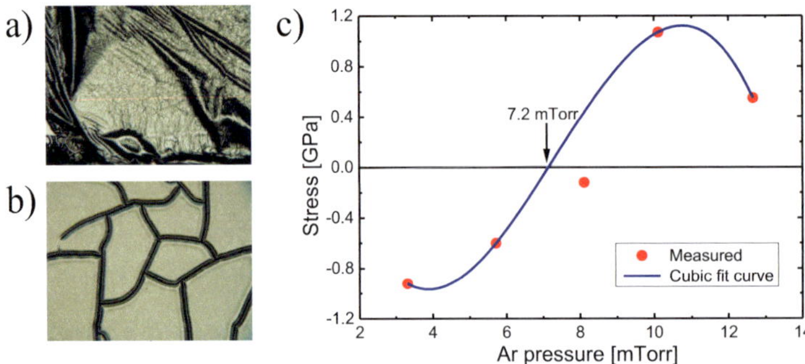

Figure 4.3: a) Nb film sputtered with $p_{\text{Ar}} = 3.25$ mTorr on photoresist. b) Nb film sputtered with $p_{\text{Ar}} = 13.3$ mTorr on photoresist. c) Nb film stress as a function of the Ar sputtering pressure.

the soft underground. In this way, the compressive stress can be seen as wrinkles in the film (see Figure 4.3a) while tensile stress leads to cracks in the film (see Figure 4.3b). With this method, minimal film stress seemed to be obtained for Ar pressures around $9 - 10$ mTorr.

However, the growth mechanisms of films vary with their substrates, so that these results cannot be directly transferred to Josephson junctions fabricated on oxidized silicon. Consequently, Nb films with thicknesses $d_F \approx 200$ nm were sputtered on 2-inch oxidized silicon wafers. In order to obtain comparable film thicknesses, the deposition rate r_{dep} was determined for each Ar pressure. The profile of each wafer was measured on a scan length $L_{scan} = 47$ mm along the same line before and after film deposition using a DEKTAK profilometer. For four out of five wafers, the difference of the two profilometer curves gave clear parabolas, so that the bending K of the wafers could be read from the data (see Table 4.2). Using the Pythagorean theorem, this bending could be transformed into the radius of curvature R_{curv} according to:

$$R_{curv} = \frac{L_{scan}^2}{8K} + \frac{K}{2}.$$
(4.1)

Knowing the radius of curvature of the wafer and several material parameters, the stress σ in the Nb film can be calculated using [70]

$$\sigma = -\frac{1}{6R_{curv}} \cdot \frac{E_Y}{1-v} \cdot \frac{d_s^2}{d_F}.$$
(4.2)

Here, $d_s = 300$ µm is the thickness, $E_Y = 169$ GPa Young's modulus and $v = 0.279$ the Poisson number of the silicon substrate. The Nb film stress as a function of the Ar pressure can be found in Table 4.2 and Figure 4.3c. The observed non-monotonous dependence is typical for sputtering processes and very similar to the one found in [51]. The measurement point at $p_{Ar} = 8.1$ mTorr, where no clear parabola could be observed, was omitted and the rest of the data was used to determine a stress-free working point at $p_{Ar} = 7.2$ mTorr by a cubic fit (which does not represent any known physical law but nicely reproduces the dependence observed in [51]). This value varies significantly from the one of $p_{Ar} = 5.5$ mTorr used before the start of this thesis, so that it can indeed be assumed that stress in the films was present and led to a reduced junction quality.

Table 4.2: Measured bending of silicon wafers under Nb film stress and deposition rates for different Ar working pressures. These values were used to calculate the Nb film stress according to (4.2). The numbers in brackets are considered to be unreliable, since no clear parabola was found in the wafer profile.

p_{Ar} [mTorr]	K [µm]	r_{dep} [nm/s]	d_F [nm]	R_{curv} [m]	σ [GPa]
3.3	13.03	0.75	181	21.2	-0.92
5.7	8.94	0.76	191	30.9	-0.60
8.1	(1.7)	0.77	185	(162.4)	(-0.12)
10.1	-16.16	0.80	192	-17.1	1.07
12.65	-8.25	0.83	191	-33.5	0.55

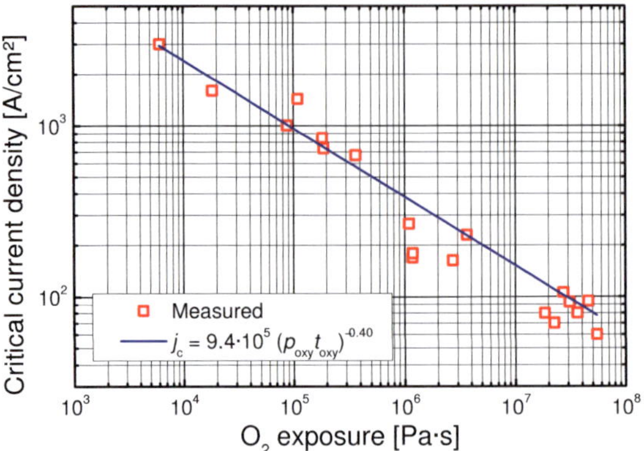

Figure 4.4: Critical current density j_c versus oxygen exposure $(p_{oxy}t_{oxy})$. The observed dependence is close to those found in literature (see text).

Formation of Tunneling Barrier

In section 4.1.1, the importance of a precise definition of the Al thickness for the junction quality was discussed. Hence, the Al sputtering rate was determined carefully to be 0.35 nm/s. Consequently, the Al sputtering time was increased from 15 s to 20 s, which should increase the Al thickness from 5 nm to the optimal value of 7 nm.

The oxidation of aluminum is carried out in the load lock of the UTS 500 vacuum system in a pure oxygen atmosphere at room temperature. Before the wafer is brought into the load lock, the latter is rinsed three times with oxygen. The oxidation usually takes place for $t_{oxy} = 30$ min, but for high critical current densities > 2 kA/cm^2, t_{oxy} can be reduced to lower values, e.g. 10 min. The critical current density is controlled by the oxygen pressure p_{oxy} during oxidation and should depend on the oxygen exposure $p_{oxy} \cdot t_{oxy}$ like [71]

$$j_c \propto (p_{oxy}t_{oxy})^{-\kappa}. \tag{4.3}$$

This dependence can be seen in Figure 4.4. The exponent $\kappa = -0.40$ found by a fit according to (4.3) is very close to the value of $\kappa = -0.38$ found in [71]. Furthermore, Figure 4.4 shows that we can control the critical current density between 50 A/cm^2 and 3 kA/cm^2.

In the past, the gas inlet and the setting of the oxidation pressure was carried out by hand using manual valves. However, the oxidation conditions should be reproducible and the desired p_{oxy} value should be reached as precisely as possible. Consequently, the oxidation procedure was automatized, so that now the user can simply set the desired pressure. Then, a micro-controller calculates the required valve opening times and gas flow settings and opens an electromagnetic valve. Details about this system and this procedure can be found in the *Studienarbeit* of Jochen Antes [Ant09].

4.2.2 Reactive-Ion-Etching

The reactive-ion-etching (RIE) process at the IMS is based on tetrafluoromethane CF$_4$ and oxygen O$_2$. Before the start of this thesis, etching was carried out at a CF$_4$ flow of

$F_1 = 29$ sccm, an O_2 flow of $F_2 = 5.9$ sccm, a pressure of $p = 260$ mTorr and a power of $P = 100$ W. These parameters resulted in an etching rate of around 560 nm/min and the occurrence of resist residue on the chip. Additionally, the resist remaining on the protected structures after etching was very hard to clean off. This phenomenon is often explained with the formation of polytetrafluoroethylene (PTFE). For more precise and cleaner etching, parameters with a slower Nb etching rate and a non-negligible resist etching rate needed to be found. Furthermore, etching with the new parameters should be reproducible and spatially homogeneous.

Since it is very hard to model plasma processes like RIE analytically, the etching properties at a certain set of parameters cannot simply be extrapolated from the properties at another set of parameters. Consequently, the parameter space has to be searched experimentally for a working point with the required etching properties. In this thesis, this search was carried out using the response surface methodology (RSM). This method as well as the detailed experimental parameters are discussed in Appendix C and in the *Studienarbeit* of Maher Rezem [Rez11].

Finally, this modeling procedure allowed to chose the following parameter set for all RIE processes in Josephson junction fabrication: CF_4 flow of $F_1 = 49$ sccm, O_2 flow of $F_2 = 21$ sccm, pressure of $p = 200$ mTorr and a power of $P = 100$ W. For these parameters, moderate etching rates $r_{e,Nb} = 88$ nm/min (for Nb) and $r_{e,res} = 81$ nm/min (for photoresist) were obtained. Optical investigation with a microscope showed very clean chip surfaces and also the resist on the remaining Nb structures was easy to clean off. Furthermore, a high spatial uniformity of the Nb etching rate was observed, as can be seen in Figure 4.5.

For more detailed information about the chosen etching process, time-resolved measurements of the etching rate were carried out. It was found that actual etching only started after a dead time t_0 and then continued more or less proportional to elapsed time $t - t_0$. If the etching plasma was operated for at least 3 min in the chamber before the chips were brought in (preconditioning), t_0 decreased significantly. Furthermore, the etching depth was much more linear with time for preconditioning with an empty chamber (see Figure 4.5) than

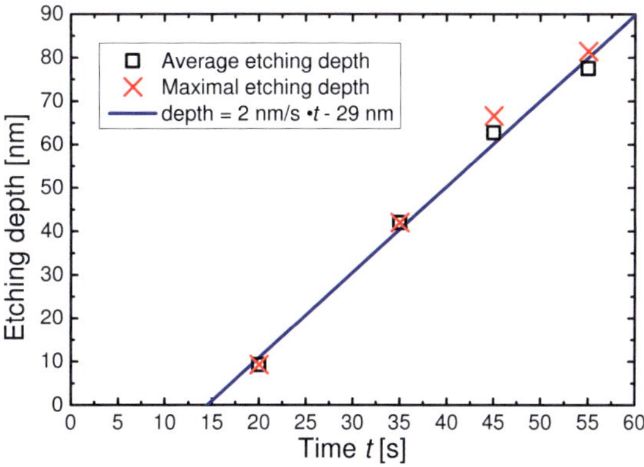

Figure 4.5: Nb etching depth versus time. Preconditioning was carried out for 3 min without Nb in the chamber.

for preconditioning with a large Nb block in it. Figure 4.5 also shows the average and the maximal etching depth on a chip. The fact that these two points differ only slightly shows the high spatial uniformity of the etching process.

4.2.3 Anodic Oxidation

Anodic oxidation is an important step in Josephson junction fabrication, as an insulating layer of Nb_2O_5 is formed directly out of the bottom Nb electrode and thereby provides ideal coverage. This is especially important at the edges of the M2a layer, since the SiO might not provide ideal insulation there. Consequently, anodic oxidation helps avoiding shortcuts, lowering subgap currents and increasing the yield. For the calculation of capacitance values, it is important to know the permittivity of Nb_2O_5. Values given in literature vary between $\varepsilon_r = 29$ and $\varepsilon_r = 40$. In various different experiments performed in this thesis, a value of $\varepsilon_r \approx 33$ was found for the Nb_2O_5 layers fabricated at the IMS.

In the fabrication process employed before the start of this thesis, parts of the M2a edges were not oxidized (see Figure 4.1). In order to change this, the order of the processing steps was changed. Now, the M2a is patterned first and only then the junctions J are defined and the anodic oxidation is carried out (see Figure 4.7). This makes the fabrication of isolated structures (such as a phase qubit, which is a stand-alone Nb ring) more laborious: In chip design, these structures have to be connected to the rest of the circuit by Nb bridges in order to ensure that they are also anodized. These bridges must be removed at the end of the fabrication, which implies an additional processing step. However, due to the expected higher junction quality, this consequence was accepted.

The anodization process is carried out in an aqueous solution of $(NH_4)B_5O_8$ and $C_2H_6O_2$ at room temperature. Per applied volt, 0.88 nm Nb are turned into 2.3 nm Nb_2O_5. The current I_{AO} in the electrolyte during the rise of the voltage V was limited in order to obtain a slow Nb_2O_5 growth. Once the set voltage level is reached, I_{AO} decreases exponentially. Anodization is stopped after a certain time t_a.

One problem in the use of anodic oxidation is the fact that Nb_2O_5 tends to creep under

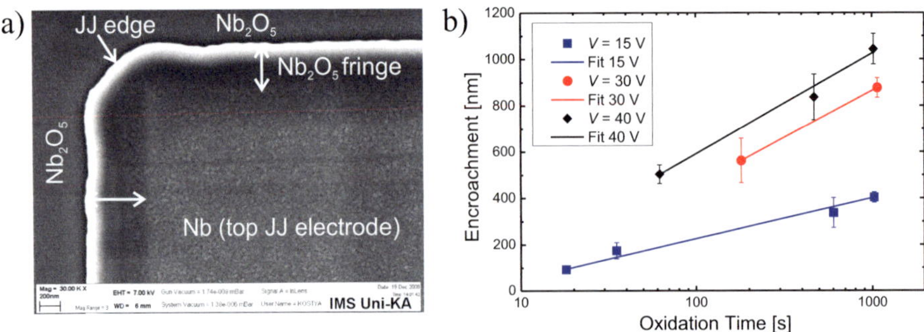

Figure 4.6: a) SEM picture illustrating how Nb_2O_5 (dark), created during anodic oxidation, has crept under the resist, forming an insulating fringe and leaving a reduced Nb (light grainy) surface. b) Dependence of the encroachment on the anodisation time t_a for different voltages.

the resist which is supposed to protect the Josephson junction surface (see Figure 4.6a). This is usually referred to as encroachment [72, 73]. For small junctions with sizes in the μm range, a closed Nb_2O_5 layer can form and hence turn them defective. For very small junctions, this problem can only be solved by the use of a hard mask, which will be discussed in section 4.3. However, problems can already occur for Josephson junctions with sizes $\geq 5 \times 5$ μm^2 defined by photolithography. We found it necessary to harden the photoresist AZ 5214 E, which is used in our fabrication, by baking it for several minutes at 120 °C after development and directly before RIE. This ensured good adhesion and minimal encroachment during the following anodization.

In order to determine the dependence of the encroachment D on the anodization parameters, we varied the latter and examined the samples under the scanning electron microscope (SEM). The proportionalities $D \propto V$ and $D \propto \ln t_a$ were found (see Figure 4.6b). On test Josephson junctions, it was found that the junction quality did not improve anymore after a certain anodization time. Hence, it seems that the end of the exponential tail of I_{AO} only comes from the fact that Nb_2O_5 creeps further under the resist. Consequently, the anodic oxidation was usually stopped after $t_a = 6$ min in Josephson junction fabrication.

4.2.4 Overview of the New Technological Process

A schematic overview of the newly developed process can be seen in Figure 4.7. The changes made with respect to the old fabrication process described in section 4.1 are given in the following list:

- The thickness of the M2a layer has been changed to 100 nm.
- A working point for minimal Nb film stress has been found.
- The thickness of the Al layer has been corrected to 7 nm.
- The bottom electrode is now patterned before the junction, so that all M2a sidewalls are anodically oxidized.
- New etching parameters for slow, clean and uniform RIE have been determined using response surface methodology.

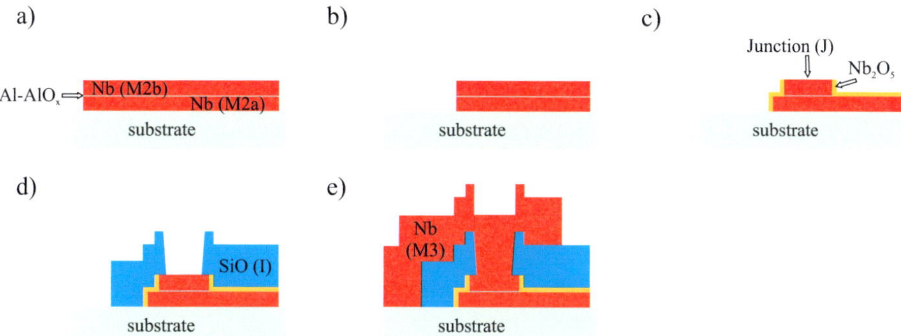

Figure 4.7: Schematic overview of the new Josephson junction fabrication process: a) As-deposited trilayer b) After patterning of bottom electrode c) After junction definition and anodic oxidation d) With deposited SiO insulation layer e) Final structure with wiring layer.

- Post-baking of the photoresist before the anodic oxidation has been introduced to avoid encroachment of the Nb_2O_5 under the resist as much as possible.
- The deposition of the M3 wiring layer has been transferred to the Univex 450 vacuum system, which will be discussed in detail in section 4.3.3.

4.3 Development of a new Process for sub-μm to μm-size Junctions

Since the junction sizes which are required for the quantum experiments performed in this thesis are smaller than 5×5 μm^2, they could not be fabricated using the standard photolithography process discussed in section 4.2. Instead, a new fabrication process needed to be developed for junctions in the sub-μm to μm range. Such small Josephson junctions are also required or advantageous for many applications such as high-speed superconducting digital circuits, programmable voltage standards, SQUIDS, mm-wave receivers and sub-millimeter wave mixers.

Considering the multitude of possible applications, it is not surprising that several fabrication techniques for sub-μm to μm-size junctions have been developed by various research groups over the years. While a precise and reproducible definition of the junction size is routinely performed with the help of electron-beam lithography, the application of anodic oxidation in such processes remains a challenge. This is due to the fact that the Nb_2O_5 layer formed during the anodic oxidation creeps under the resist mask protecting the top electrode of the junction (see Figure 4.6). Even for a plasma hardened resist and an anodization voltage of 20 V, a significant encroachment was still observed [72]. Some groups solve this problem by simply omitting the anodic oxidation or performing it at very low voltages below 10 V [74–77]. Since for the quantum measurements performed in this thesis, very low subgap leakage and hence sufficiently thick Nb_2O_5 insulation layers are needed, this was not an option. Other research groups have replaced the resist mask with an insulating hard mask such as SiO or SiO_2 [73, 78–81], which entirely eliminates the encroachment. These hard masks have to be removed by an additional process step, either by reactive ion etching (RIE) through the insulator right above the Josephson junction or by chemical mechanical polishing (CMP). However, it has been found that plasma processes [82] as well as CMP [83] can damage the Josephson junctions. Furthermore, it is being discussed that such processes might degrade the Nb surfaces and lead to the formation of spurious two-level systems (TLS), which limit the coherence times in superconducting phase qubits [23, 84]. Consequently, a novel method using an aluminum hard mask, which can be removed by a fast wet etching process with KOH, was developed in this thesis. The wet etching does not alter the Nb surfaces and should not damage the Josephson junctions.

In general, the fabrication process for sub-μm to μm-size Josephson junctions is still performed as shown in Figure 4.7 and discussed in section 4.2. However, three steps of the standard photolithography process are modified:

- The Josephson junctions are patterned with the help of electron-beam lithography and an Al hard mask (see section 4.3.1).
- The SiO insulating layer is defined by electron-beam lithography and lift-off (see section 4.3.2).

- An *in-situ* pre-cleaning is performed before the deposition of the M3 wiring layer (see section 4.3.3).

4.3.1 Definition of Josephson Junctions

At first, it was tried to use the negative e-beam resists AR-N 7500 and AR-N 7520 for the definition of the Josephson junctions. But analogous to the usage of photoresist and the findings reported in literature (see above), a Nb_2O_5 encroachment under the resist of at least $\gtrsim 500$ nm was found (see Figure 4.8a). Consequently, no electrical contact from the wiring M3 to the bottom electrode M2b was possible for smaller junctions. In fact, the smallest junction which could be fabricated in this way had a diameter of $d = 1.35$ μm; but already for such sizes an electrical contact was not reached reliably. Furthermore, patterning with the negative e-beam resist was not very reproducible. The obtained junction size varied from process to process and had to be measured by SEM each time. For applications which require the precise definition of junctions of a certain size, patterning with e-beam resist was not an option even for larger junction sizes with diameters around 3 μm. An example of different obtained junction sizes for a design value of 3 μm can be found in Table 8.1.

In order to avoid encroachment, facilitate the fabrication of sub-μm junctions and obtain a high process-to-process reproducibility, we developed a hard mask technique similar to the ones used in literature (see above). However, we wanted to avoid any further processes which might damage the junctions or alter the Nb surfaces such as plasma etching or CMP. Hence, insulators like SiO or SiO_2 were ruled out as hard mask material. To find an alternative, we searched for a metal that could be anodized, so that it would not act as a shortcut for the anodization current, and that could be removed selectively by a clean wet etching process. Since aluminum fulfills all of these criteria, it was chosen as the hard mask material. The mask is created by employing positive e-beam lithography with a 235 nm thick resist and carefully chosen e-beam parameters (for more details see Table B.2). Afterwards, 50 nm of Al are DC magnetron sputtered in the UTS 500 vacuum system and subsequently lifted off.

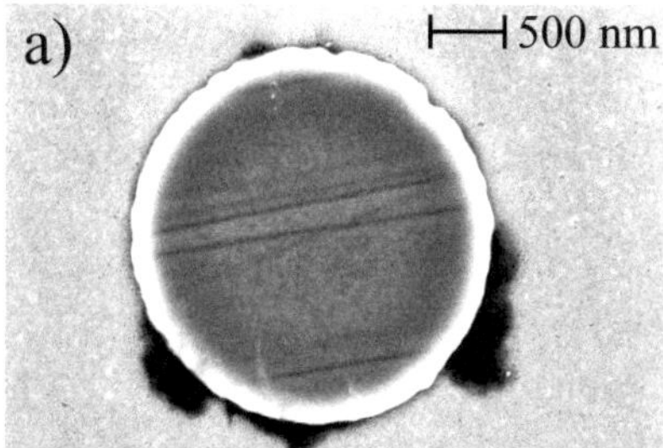

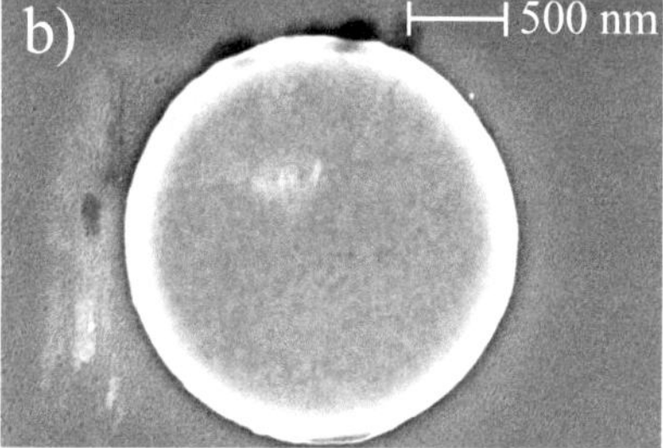

Figure 4.8: a) Micrograph of Josephson junctions patterned with negative e-beam resist. A rim with ≈ 500 nm width without Nb grains can be seen, which is due to encroachment. Furthermore, residue of the negative e-beam resist can be seen (dark areas). b) Micrograph of Josephson junctions patterned with an Al hard mask. No encroachment is observed and the Nb grains can bee seen up to the edge of the junction.

This Al mask is then used for patterning of the junctions, which begins with etching the top electrode of the trilayer by RIE. Since aluminum is not etched here, both the AlO_x tunneling barrier and the Al mask act as ideal etch stoppers. Now, the Nb_2O_5 layer is formed by anodic oxidation as described in section 4.2.3. Per applied volt, 0.88 nm Nb are turned into 2.3 nm Nb_2O_5 and 0.9 nm Al are turned into 1.3 nm AlO_x [85], so that for a typical anodization voltage of 25 V, only half of the Al hard mask thickness is oxidized and the underlying Nb is protected. The hard mask is removed by wet etching with KOH, which does not alter the Nb surface and should not damage the Josephson junctions. Furthermore, the junction size defined with the optimal lithography parameters was measured with SEM, found to be identical to the design value and did not vary from process to process anymore (this can also be seen in Figure 5.9).

4.3.2 Formation of SiO Vias

In the Josephson junction fabrication process, a dielectric layer is needed to insulate the bottom M2a electrode from the M3 wiring layer and realize overlay capacitors. The material chosen for this insulating layer should fulfill several requirements. First, it should exhibit good insulation properties, i.e. a low density of defects and pinholes as well as a good coverage of the underlying Nb edges. Second, it should be possible to deposit it at rather low temperatures, since this layer is patterned by a lift-off process. This is especially important for the sensitive e-beam resist used in the fabrication of sub-μm junctions. Additionally, many Josephson junction based applications require low dielectric losses in the circuits. For superconducting phase qubits, which were fabricated within this thesis, high losses can even prevent their general operability (a more detailed discussion of dielectric losses can be found in chapter 6 while a discussion of decoherence mechanisms in phase qubits is given in chapter 9).

At the IMS, two systems for the formation of insulating layers were available at the time of this thesis: A thermal evaporation system for SiO and an RF sputtering system for SiO_2. Both materials were investigated considering their usability for Josephson junction fabrication. It was found that a clean lift-off of SiO_2 layers was only possible for low working pressures during sputtering. However, this parameter regime exhibited very high dielectric losses [86]. But even in a systematic investigation of different working pressures

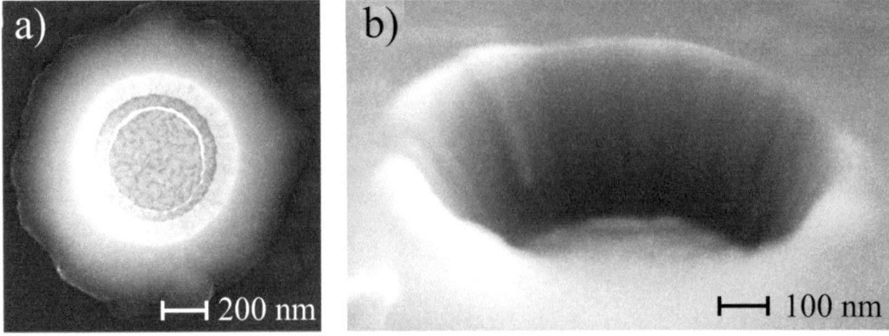

Figure 4.9: Micrograph of SiO vias in top view (a) and diagonal perspective (b).

and combinations of various process gases, the dielectric losses in the RF sputtered SiO_2 were always larger than in the thermally evaporated SiO (this is discussed in more detail in section 6.5.1). Consequently, SiO was chosen for Josephson junction fabrication. The literature value of the permittivity of SiO is $\varepsilon_r = 5.7$, which was confirmed for samples fabricated at the IMS. This means that the material deposited by thermal evaporation in the high vacuum chamber is probably indeed SiO and not some undefined silicon oxide SiO_x.

In order to avoid leakage currents at the edges of the Josephson junction, the SiO via on top should always be a few hundred nanometers smaller than the junction itself. This means that vias for sub-µm junctions need to have a diameter of 400 to 500 nm or even smaller. Since the resist should be thicker than the SiO layer (typically 250 to 300 nm) to facilitate a reliable lift-off, the aspect ratios of the resist islands on top of the junctions reach values as small as ≈ 1. These resist islands must be defined by e-beam lithography and have to be stable under SiO evaporation until the lift-off has been successfully performed. For this technological challenge, two main ingredients have been found to be crucial.

First, a suitable resist thickness has to be found and the e-beam lithography parameters have to be optimized carefully. We obtained the best results with a resist of 465 nm thickness; the detailed e-beam parameters can be found in Table B.2. Second, we found that an increase of the sample temperature during thermal evaporation does not permit the fabrication of small vias. In order to monitor the sample temperature, a thermocouple was glued onto the sample holder with silver conductive paint. Although the sample holder was water-cooled, temperatures beyond 40 °C were observed during evaporation. As a first remedy, a radiation shield was installed between the crucible and the sample holder, which decreased the maximal temperature slightly. However, the main impact was reached by cleaning the cooling water loop and changing its design. Before, it cooled the oscillating crystal (used for *in-situ* film thickness measurements), the crucible electrodes and the sample holder in series. This was changed to parallel cooling of all the above components, which reduced the maximally observed temperature during evaporation to ≈ 23 °C. The evaporation of SiO took place at pressures of $< 10^{-6}$ mbar and deposition rates of around 40 nm/min.

As can be seen in Figure 4.9, vias with diameters as small as ≈ 140 nm could be fabricated. However, this processing step remains the only one which still limits the yield of the process for sub-µm to µm-size junctions due to incomplete lift-off. In the fabrication of 16 Josephson junction chains containing 20 junctions each (see also section 5.4.3), it was found that in average, two vias were not lifted off per chain. This is equivalent to a yield of $18/20 = 90$ %. Furthermore, one junction with a not lifted-off via prevents measurement of the entire chain. We found that 3 out of 16 chains were working, which corresponds to a yield of $(3/16)^{1/20} = 92$ %. These values are not sufficient for using this process for RSFQ or voltage-standard applications involving thousands of junctions. But for the fabrication of samples for quantum experiments as performed in this thesis, such a yield is largely sufficient. The chains contained junctions with diameters of 1, 2 and 4 µm. It can be assumed that for ultra-small vias with diameters below 500 nm, the yield is somewhat lower. A possibility to increase the yield of the SiO via lift-off might be to employ more sophisticated e-beam resists, such as the double-layer resists routinely used in the Al shadow evaporation technique.

4.3.3 Deposition of Wiring Layer

Just as in the standard junction fabrication process described in section 4.2, the wiring layer M3 is also defined by negative photolithography for the sub-μm to μm-size junctions. At fist, the Nb deposition was also carried out by DC magnetron sputtering in the main chamber of the UTS 500 system with the standard RF Ar plasma pre-cleaning in the load lock of the system. However, it was found that no electrical contact between the M3 and the M2b layers could be reached through the small vias required for sub-μm Josephson junctions. Consequently, the M3 deposition was transferred to the Univex 450 vacuum system, where an ion gun was available for *in-situ* pre-cleaning. A systematic study with different pre-cleaning parameters showed that a reliable electrical contact could already be reached for low-energy Ar ions at a beam voltage of 100 V and a beam current of 10 mA. Pre-cleaning at a beam voltage of 300 V and a beam current of 30 mA also led to a good electrical contact, but resulted in Josephson junctions which mostly had increased subgap leakage currents, excess currents or were even defective. This can be attributed to the higher energies of the Ar ions, which probably penetrated the M2a electrode and created pinholes in the tunneling oxide. Consequently, the pre-cleaning parameters of lower energy were chosen for Josephson junction processing. Since no decrease in junction quality was observed, this process was also taken over for the M3 deposition of standard photolithography junctions.

5 Characterization of Junction Quality

The fabrication processes that have been developed and discussed in chapter 4 were used to fabricate Josephson junctions of various sizes and critical current densities. In this chapter, the quality parameters discussed in chapter 3 are employed to evaluate their quality. Since a new measurement system had to be set up in order to be able to characterize the Josephson junctions, this is discussed in some detail first. After that, the quality of the Nb films used in the trilayers is evaluated. Thereafter, Josephson junctions fabricated with the optimized standard photolithography process are characterized. Finally, the quality of the sub-µm to µm-size junctions fabricated with the newly developed fabrication process discussed in section 4.3 is determined.

5.1 New Measurement Setup at the IMS

Before the start of this thesis, the measurement system used for Josephson junction characterization did not allow measurements of reasonable quality. For underdamped junctions, the critical currents I_c were strongly suppressed and the obtained Fraunhofer modulations were very noisy, even for rather large I_c values of a few hundred µA. It was concluded that this was due to insufficient filtering and shielding. Consequently, a new measurement setup was built, including a new PC, a new data acquisition (DAQ) card, a new dipstick including

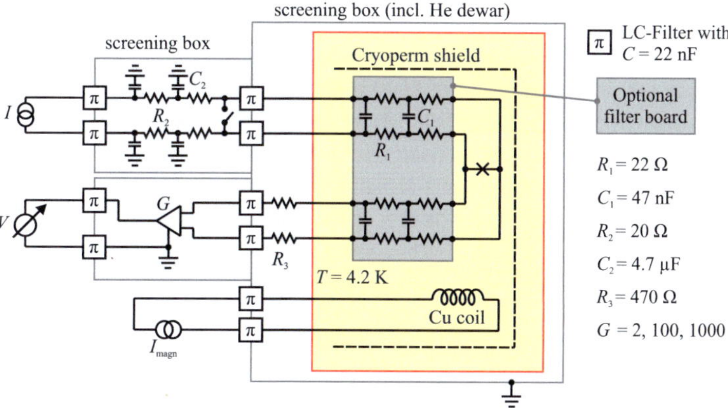

Figure 5.1: Measurement setup for Josephson junction characterization at the IMS. Inside the *Cryoperm* shield, a copper coil with about 9200 windings can create a magnetic field through the junctions. The resistors R_3 only serve to stabilize the amplifiers. The switch can be used to isolate the junction from the measurement system, in order to save it from potential voltage pulses due to the plugging of cables.

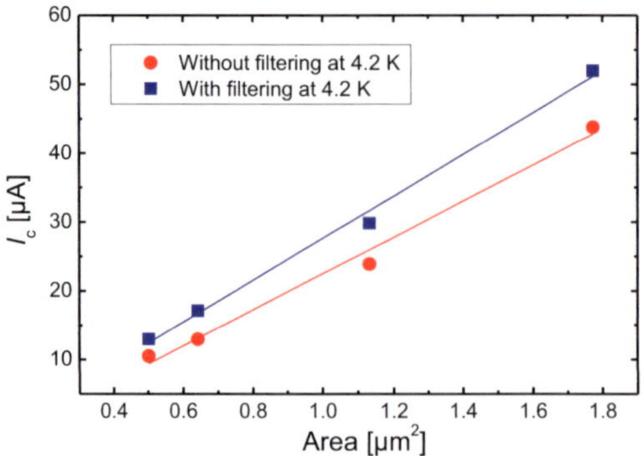

Figure 5.2: Observed critical currents with and without additional filtering stage for Josephson junctions with $j_\mathrm{c} = 3$ kA/cm². A clear improvement can be seen when filtering at $T = 4.2$ K is added.

built-in amplifiers and filters, a new magnetic shield and new magnet coils. An overview of the new setup can be seen in Figure 5.1.

As DAQ card, the model PCI-6052E from *National Instruments* was chosen. It contains digital-to-analogue converters (DAC) and analogue-to-digital converters (ADC) with a resolution of 16 bit and has a maximal sampling rate of 333 kS/s. The software *GoldExi*[1] running on the PC controls the DAC output voltage up to 10 V, which is converted into a current by an analogue transconductance amplifier with amplifications 10^{-2} A/V, 10^{-3} A/V, 10^{-4} A/V and 10^{-5} A/V. This current is fed through the room temperature filtering stages depicted in Figure 5.1. Even with this new setup, we found that the critical currents were still lower than expected. However, by employing an additional filtering stage at $T = 4.2$ K, the expected I_c values could be reached. The significant increase in critical current (of about a factor of 1.2) due to the supplementary filtering stage can be seen in Figure 5.2. The reason for this is that without the additional stage, the noise created by the current source and the filtering resistors at room temperature directly influences the Josephson junctions. With added low-temperature filtering, this noise is dissipated by the resistors at $T = 4.2$ K. Instead, the junctions are only influenced by the noise created by the filtering resistors at $T = 4.2$ K. The bandwidth of the current path can be estimated as $f = 1/(2\pi R_\Sigma C_\Sigma)$ with R_Σ and C_Σ being the sum of all resistances and capacitances, respectively. This results in $f = 50$ Hz, so that all radio frequency signals and even parasitic signals coming from mains electricity should be filtered. This bandwidth also results in a reasonable measurement speed. However, for high resistive samples with $R > 1$ kΩ, the bandwidth is reduced to below $f = 10$ Hz, which requires relatively slow measurements.

The voltage over the sample is also *RC* filtered and then amplified by an instrumentation amplifier. The latter is built into the dipstick and its gain can be adjusted to $G = 2$, 100 and 1000. Its design was taken over from the one used in the dipstick of the *Messknecht* system at the IMS. Finally, the voltage is fed to the analogue-to-digital converter of the data acquisition card and recorded by the *GoldExi* software.

---

[1] *GoldExi* is a software developed by Edward Goldobin from the Universität Tübingen. More information can be found at http://www.reocities.com/goldexi/.

In order to measure $I_c(\Phi)$ modulations, we made a copper coil with about 9200 windings at the Institut für Technische Physik (ITEP). The current through this coil is also controlled by the *GoldExi* software, a second DAC and a second transconductance amplifier.

5.2 Results for Nb films

Before using the Nb films deposited in the UTS 500 vacuum system for actual Josephson junction trilayers, their film quality was evaluated. For this, the films were sputtered on oxidized silicon with the parameters for minimal film stress found in section 4.2.1. One common measure for the film quality is the critical temperature T_c. Although Nb cannot be described in the weak-coupling limit (see section 3.1.1), the critical temperature is still proportional to $2\Delta(0)$ and hence a good measure of film quality. The second common measure for Nb film quality is the residual resistance ratio (RRR). The latter gives information about the purity of the films.

According to Matthiessen's Rule, the resistivity in a metal is given by

$$\rho(T) = \rho_0 + \rho_{Ph}(T),\tag{5.1}$$

where ρ_0 is the resistivity due to electron scattering on crystal defects and $\rho_{Ph}(T)$ the resistivity due to electron scattering on phonons. Since phonons are completely frozen out at $T = 10$ K, the residual resistance ratio

$$\mathrm{RRR} = \frac{R(300\,\mathrm{K})}{R(10\,\mathrm{K})} \approx \frac{\rho_{Ph}}{\rho_0}\tag{5.2}$$

gives a measure for the defect density in a metal. The higher the RRR value, the fewer defects contribute to electron scattering.

The Nb films were characterized in the *Messknecht* measurement system at the IMS. For a 250 nm thick film, a RRR value of 4.1 and a transition temperature of $T_c = 9.2$ K

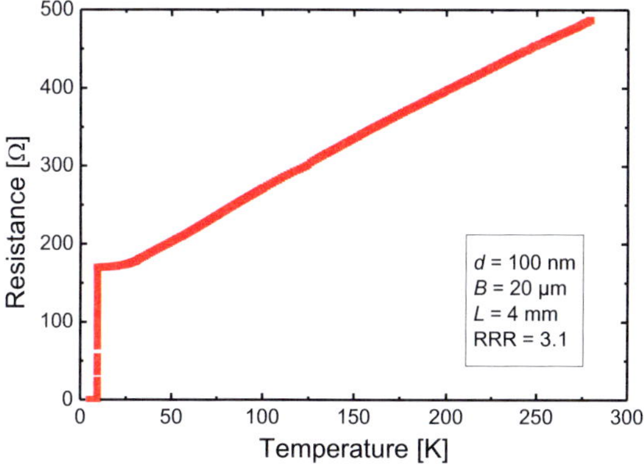

Figure 5.3: RRR measurement on a 100 nm thick Nb film deposited with the parameters for minimal film stress discussed in section 4.2.1.

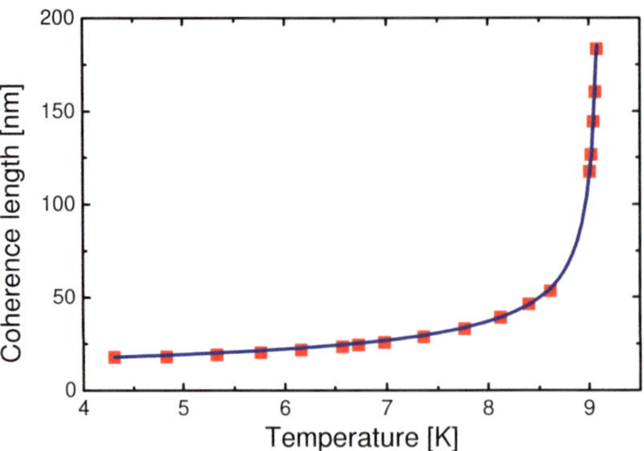

Figure 5.4: Coherence length determined by (5.3) versus temperature for a 100 nm thin Nb film deposited with parameters for minimal stress in the UTS 500. This film was deposited without RF pre-cleaning. The blue line was obtained by a fit with (2.6), giving $T_c = 9.11$ K and $\sqrt{\xi_0 \ell} = 15.1$ nm.

were obtained. On films with the same thickness as actually used for the Josephson junction trilayers (100 nm), a RRR value of 3.1 and a critical temperature of $T_c = 9.2$ K were observed (see Figure 5.3). From the geometry of a 100 nm thick Nb bridge (length of 4 mm, width of 20 µm), the resistivity was calculated to be $\rho_0 = 8.5 \cdot 10^{-8}\,\Omega\text{m}$ and $\rho(300\,\text{K}) = 2.6 \cdot 10^{-7}\,\Omega\text{m}$. These values confirm the expected high quality for stress-free Nb thin films. Furthermore, the Nb films showed excellent microwave properties and were used for the fabrication of resonators with high quality factors [87, 88].

Furthermore, the coherence length in the Nb films was determined. This is especially important for Josephson junctions realized as weak links [89], such as Dayem bridges [25, 89], where the bridge dimensions have to be smaller or in the range of the Ginzburg-Landau coherence length $\xi(T)$. More details about Dayem bridges and their possible applications in nanoSQUIDs can be found in the *Studienarbeiten* of Max Bauer [Bau10] and Michael Merker [Mer10]. In the context of Nb/Al-AlO$_x$/Nb Josephson junctions, the coherence length is simply taken as a measure of the Nb film quality. Hence, $\xi(T)$ was determined for 100 nm thin Nb films sputtered in the UTS 500 vacuum system, just as they are used in Josephson junction fabrication. This was done by measuring the second critical magnetic field B_{c2} and using the relation [24]

$$\xi(T) = \sqrt{\frac{\Phi_0}{2\pi B_{c2}}}.$$ (5.3)

These measurements were carried out in the pulse tube cooler at the IMS. The results for a 100 nm thick Nb film can be seen in Figure 5.4. At $T = 4.2$ K, a value of $\xi(T) = 17.8$ nm was obtained, which is a rather large coherence length for Nb deposited without heating of the substrate. More details about coherence length measurements can be found in the *Studienarbeit* of Michael Merker [Mer10].

5.3 Results for Standard Junctions

Josephson junctions which were fabricated with the optimized photolithography process described in section 4.2 were characterized at $T = 4.2$ K in the newly built measurement system (see section 5.1). The quality parameters extracted from the *IV* characteristics indicate a very high quality, as will be discussed in detail in the following. An exemplary *IV* curve can be seen in Figure 5.5.

The high quality of the Nb trilayer electrodes (M2a and M2b), which was found in section 5.2, is also confirmed by the observed high gap voltages in the current-voltage characteristics of our Josephson junctions. They are typically above $V_{gap} \geq 2.8$ mV and go up to $V_{gap} = 2.95$ mV [Mec09], which corresponds to the theoretically expected value calculated in (3.3).

The $I_c R_N$ products are typically higher than 1.7 mV and reach values as high as 1.88 mV. Since we practically never observe excess currents in our junctions, the high $I_c R_N$ products show that we have strong Cooper pair tunneling close to the expected Ambegaokar-Baratoff behavior. This is a clear increase in $I_c R_N$ compared to the junctions fabricated before the beginning of this thesis (see section 4.1 and Table 4.1). Although we observed slightly higher gap voltages than before and the superconducting gap $2\Delta(T)$ enters the $I_c R_N$ product according to (3.4), this cannot explain the observed increase. Instead, we attribute the rise in $I_c R_N$ to the change of Al thickness in the trilayer from 5 to 7 nm, which should smoothen the surface of the underlying Nb and hence lead to a more homogeneous tunneling barrier (for details see section 4.1.1).

Furthermore, the Josephson junctions exhibit very low leakage currents. This is shown by the measured R_{sg}/R_N ratios, which are typically above 20 and not seldom reach higher values up to 40. This means that the characteristic voltages $V_m = I_c R_{sg}$ are typically larger than 40 mV and go up to around 80 mV (see Figure 5.5). Hence, all subgap quality parameters are clearly increased compared to before the start of this thesis (see Table 4.1). We attribute this to the process improvements discussed in detail in chapter 4, in particular to the fact that the M2a sidewalls are now also anodically oxidized and to the reduction of stress in the Nb films.

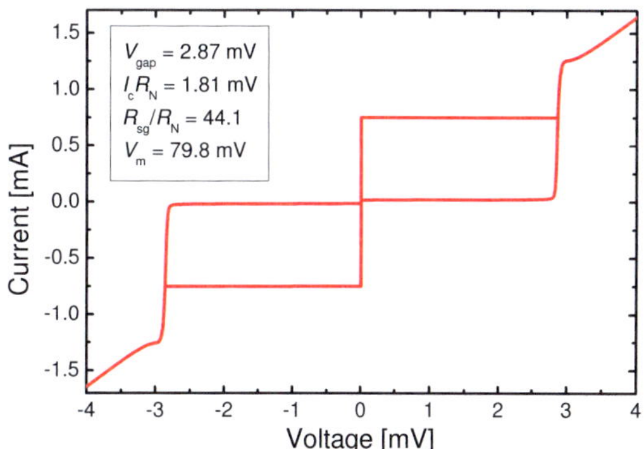

Figure 5.5: *IV* curve of a $20 \times 20\ \mu m^2$ Josephson junction with a critical current density of $j_c = 190\,A/cm^2$, measured at 4.2 Kelvin.

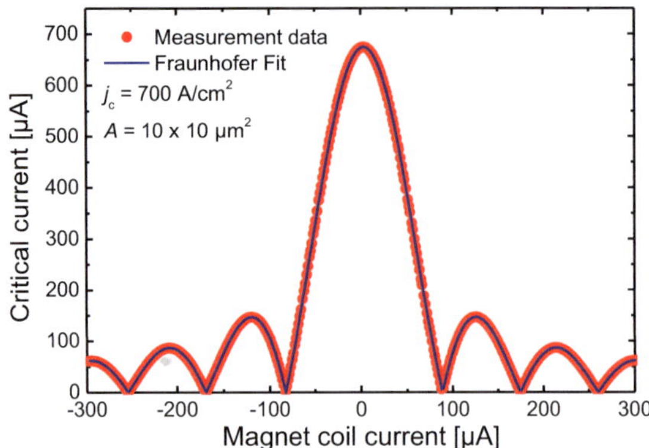

Figure 5.6: $I_{\mathrm{c}}(\Phi)$ modulation of a $10 \times 10\ \mu\mathrm{m}^2$ Josephson junction with a critical current density of $j_{\mathrm{c}} = 700\,\mathrm{A/cm}^2$, measured at 4.2 Kelvin.

We also measured $I_{\mathrm{c}}(\Phi)$ modulations and compared them to the expected behavior discussed in section 3.2. An example for a junction of area $10 \times 10\ \mu\mathrm{m}^2$ is shown in Figure 5.6. It can be seen that the measured curve is in perfect agreement with the theoretically expected Fraunhofer pattern (3.13). This shows that the critical current in our junctions is distributed absolutely homogeneously, which is only possible for a spatially homogeneous tunneling barrier.

Since Josephson junctions smaller than $10 \times 10\ \mu\mathrm{m}^2$ could not be fabricated successfully at all with the old fabrication process (see Table 4.1), we also looked if we could now fabricate $5 \times 5\ \mu\mathrm{m}^2$ junctions of good quality. An *IV* curve of such a junction can be seen in Figure 5.7. At the gap voltage, a negative differential resistance is observed, which can be attributed to local self-heating of the junctions for higher bias currents. This is rather due

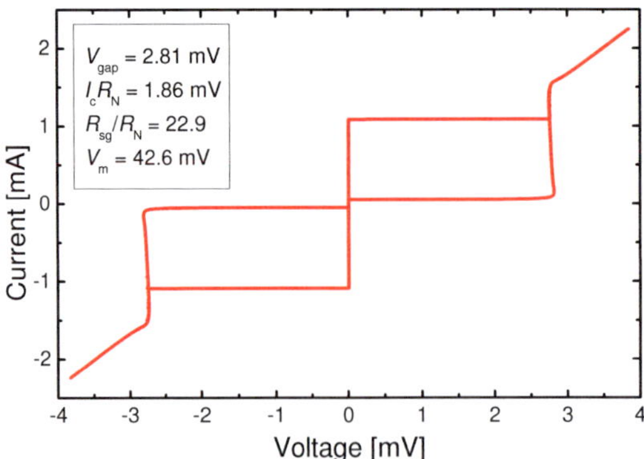

Figure 5.7: *IV* curve of a $5 \times 5\ \mu\mathrm{m}^2$ Josephson junction with a critical current density of $j_{\mathrm{c}} = 4.35\,\mathrm{kA/cm}^2$, measured at 4.2 Kelvin. The negative differential resistance at the gap voltage can be attributed to self-heating at higher current bias.

to the high critical current density $j_c = 4.35\,\text{kA/cm}^2$ than to the junction size. Apart from this effect, all quality parameters are in the range discussed above. We see that we can now fabricate $5 \times 5\ \mu\text{m}^2$ Josephson junctions without any significant decrease in quality.

5.3.1 Determination of I_c spread and yield

So far, the quality of single Josephson junctions has been discussed. In order to obtain information about properties of the fabrication process important for scaling, such as yield and parameter spread, we fabricated chains of Josephson junctions. The sample which will be discussed here contained 24 chains, of which one could not be measured due to problems with the measurement system. The chains consisted of 100 or 101 Josephson junctions each, which had either a size of $7 \times 7\ \mu\text{m}^2$ or $10 \times 10\ \mu\text{m}^2$. Altogether, 15 chains containing a total of 1508 $10 \times 10\ \mu\text{m}^2$ Josephson junctions and 8 chains containing a total of 803 $7 \times 7\ \mu\text{m}^2$ junctions were characterized.

The *IV* curve of a chain consisting of 101 $7 \times 7\ \mu\text{m}^2$ junctions can be seen in Figure 5.8a. At first sight, it looks like the *IV* curve of a single junction, but actually consists of 101 single voltage jumps of amplitude V_{gap}, so that a total voltage of $101 \cdot V_{\text{gap}}$ is reached at the vertical current rise. The individual voltage jumps occur at the 101 different I_c values of the single junctions. This means that statistics about the individual I_c and V_{gap} values can be obtained from such a measurement. However, the analysis is quite subtle, since for several junctions having the same I_c, jumps of multiples of V_{gap} can occur, which must then be attributed to a certain number of junctions. This can be difficult, since the individual V_{gap} values are unknown. Here, the *IV* curves were automatically analyzed by a *Matlab*[2] script which was carefully optimized until reliable results were reached.

In the *IV* curve shown in Figure 5.8a, nearly all junctions have very similar I_c values,

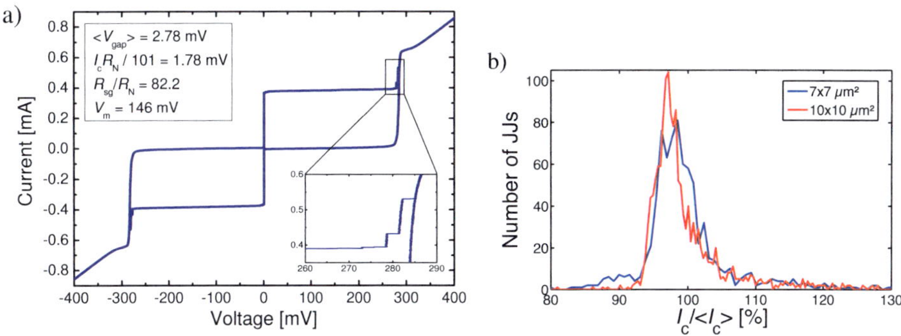

Figure 5.8: a) *IV* curve of a chain of 101 $7 \times 7\ \mu\text{m}^2$ Josephson junctions with a critical current density of $j_c = 790\,\text{A/cm}^2$, measured at 4.2 Kelvin. The inset shows abnormal junctions, which are typically present in such chains. b) Distribution of critical currents for 1508 $10 \times 10\ \mu\text{m}^2$ junctions and 803 $7 \times 7\ \mu\text{m}^2$ junctions. The highest I_c value of an abnormal junction was found at 197 % of $\langle I_c \rangle$ in a 10×10 chain.

---

[2]MathWorks, 3 Apple Hill Drive Natick, MA 01760-2098, USA

except for two junctions exhibiting significantly higher critical currents. This is a common problem and it is being discussed that the occurrence of these abnormal junctions has its origin in plasma processes [82]. Since the focus of this thesis lies on the fabrication of high-quality quantum circuits containing a rather low number of junctions, this phenomenon shall not be discussed in detail here. However, it should be mentioned that these abnormal junctions are not excluded from the following analysis.

Quality parameters were also obtained from the chain measurements. This is important, as it confirms that the high Josephson junction quality discussed so far is not only obtained on single, exceptional junctions, but actually for all junctions on a chip. Since the junctions are all connected in series, a resistance of $101 \cdot R_N$ is observed in the ohmic regime. The subgap resistance is evaluated at $101 \cdot 2$ mV and the $I_c R_N$ product is obtained as the mean of all critical currents $\langle I_c \rangle$ times the resistance in the ohmic regime divided by the number of junctions, in this case 101. It can be seen that the chain in Figure 5.8a has excellent quality parameters. Here, the R_{sg} values might be slightly overestimated due to the limited bandwidth of the measurement system and the high resistance of the chain (see section 5.1).

The analysis described so far was extended to the entire chip, so that 3211 junctions were analyzed in total. The 7×7 chains and the 10×10 chains were analyzed separately, and the results can be seen in Table 5.1. It shows that the quality parameters given previously, $I_c R_N > 1.7$ mV, $R_{sg}/R_N > 20$, $V_{gap} > 2.8$ mV and $V_m > 40$ mV, are indeed lower boundaries and easily exceeded in the average of such a large number of junctions.

Furthermore, the table shows that the yield of the optimized process accounts for ≈ 99 %. The missing 1 % of the junctions had either higher critical currents than the maximal measurement current and were hence not observed, or were not recognized by the analysis algorithm. The distribution of the critical currents can be seen in Figure 5.8b. These histograms look quite similar for 7×7 μm^2 and 10×10 μm^2 junctions, except for the little foot on the left side of the main peak for the 7×7 μm^2 Josephson junctions, which might be due to a less precisely defined junction size. The standard deviation calculated from all observed junctions accounts for $\sigma_{dev} = 7.8$ % in both cases. This value is significantly influenced by the abnormal junctions and is a measure for the entire process including the plasma processes. In order to get an idea about the j_c variation of the trilayer, the standard deviation without junctions having critical currents above $1.2\langle I_c \rangle$ was evaluated, which lies around 5 %. For high-j_c fabrication processes, which are optimized for high yield and low parameter spread, standard deviations as low as 2 % have been reached [90]. However, our

Table 5.1: Analysis of 23 Josephson junction chains on a chip with a critical current density of $j_c = 790$ A/cm^2.

	7×7 μm^2	10×10 μm^2
Yield [%]	98.5	99.3
Standard deviation σ_{dev} [%]	7.8	7.8
σ_{dev} without abnormal junctions [%]	5.5	4.9
$\langle V_{gap} \rangle$ [mV]	2.80	2.81
$\langle R_{sg}/R_N \rangle$	44.9	33.0
$\langle I_c R_N \rangle$ [mV]	1.79	1.82
$\langle V_m \rangle$ [mV]	80.4	60.0

fabrication process has been optimized for lower critical current densities and for especially high quality for quantum experiments, so that the attained yields and parameter spreads are acceptable. They should even make it possible to fabricate larger scale applications with this process.

5.3.2 Conclusions

In summary, we conclude that all aspects of the standard photolithography process have been clearly improved in comparison to before the start of this thesis. The obtained quality is excellent and should allow to carry out the desired quantum experiments. It can be concluded that the technological changes discussed in chapter 4 were correct and indeed effective. Especially, the low subgap leakage current can be attributed to the reduced Nb film stress and the anodic oxidation of the lower electrode sidewalls. The improved $I_c R_N$ products might be due to the changed Al thickness as well as the reduced Nb stress.

5.4 Results for sub-μm to μm-size Junctions

Josephson junctions with sizes in the sub-μm to μm range were fabricated with the new technological process described in section 4.3 and characterized at 4.2 Kelvin. For the entire range of critical current densities shown in Figure 4.4, this R_{sg}/R_N ratio was typically larger then 30, which indicates a very high quality. In the following, the further discussion of junction quality will be carried out on two exemplary critical current density values.

For a trilayer with $j_c = 660$ A/cm^2, we systematically varied the JJ size between 0.8 and 20 μm^2 in order to see if the Josephson junction geometries were defined precisely. Since critical currents $I_c < 10$ μA are strongly suppressed at $T = 4.2$ K due to thermal noise, the normal resistance was taken as the characteristic junction parameter. Figure 5.9 shows the measured R_N values versus the designed junction areas A. The data were fit by

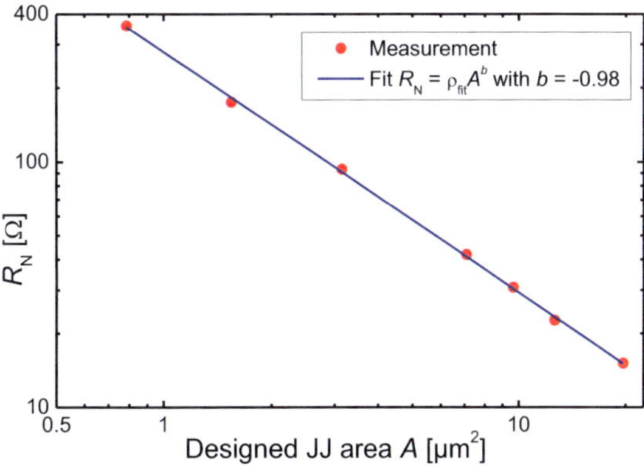

Figure 5.9: Normal resistance R_N over designed JJ area A for various circular junctions having a critical current density of $j_c = 660$ A/cm^2, measured at 4.2 Kelvin. The fit shows an excellent agreement with the expected behavior $R_N \propto 1/A$.

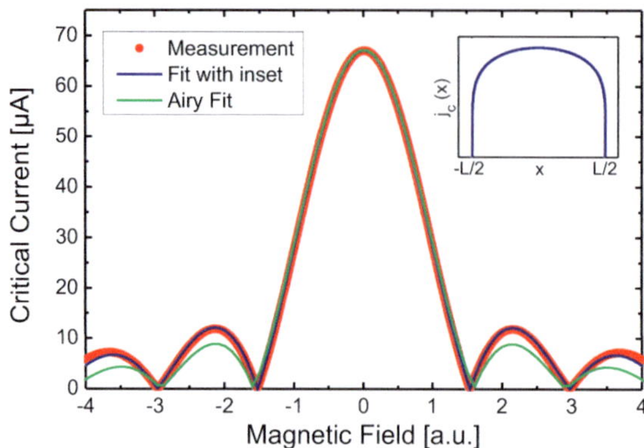

Figure 5.10: $I_{\rm c}(\Phi)$ modulation of sample B3 , recorded at $T = 20\,$mK. The blue line is the curve that was obtained as the Fourier transform of the integral critical current density distribution $\Upsilon_c(x)$ shown in the inset. The green curve was obtained by a fit with (3.14).

the power law $R_{\rm N} = \rho_{\rm fit} \cdot A^b$. It was found that $b = -0.98$, so that an excellent agreement with the expected behavior $R_{\rm N} \propto 1/A$ was obtained, which shows that the junction sizes are accurately defined by the Al hard mask process. This was confirmed by occasional measurements of the JJ size by scanning electron microscopy (SEM). Furthermore, the fit value of $\rho_{\rm fit} = 279\ \Omega\mu{\rm m}^2$ leads to an average value of $I_{\rm c}R_{\rm N} = j_c\rho_{\rm fit} = 1.84$ mV, showing the strong Cooper pair tunneling close to the expected Ambegaokar-Baratoff behavior (see section 3.1.2).

In the scope of the macroscopic quantum tunneling experiments performed at the Institut für Festkörperphysik (IFP) described in chapter 7, four of these samples with diameters from 1.9 to 3.8 µm were characterized at $T = 20$ mK, where the critical current of the junctions is not suppressed (an exemplary *IV* curve can be seen in figure 5.13). Here, the

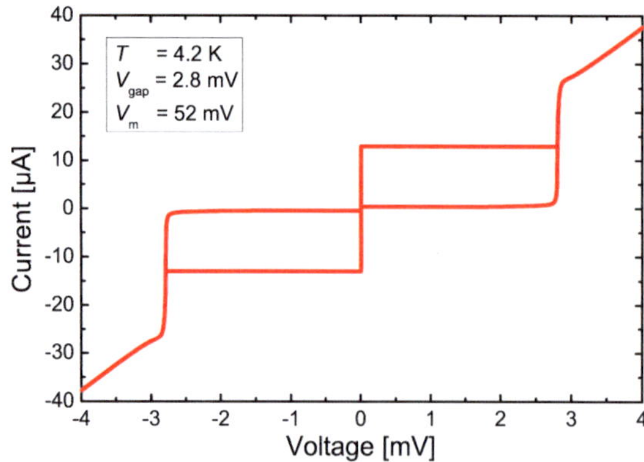

Figure 5.11: *IV* curve of a circular junction with an area of 0.5 µm^2 and a critical current density of $j_{\rm c} = 3\,$kA/cm^2, measured at 4.2 Kelvin.

gap voltages and the $I_c R_N$ products given in Table 5.2 confirm the values obtained at 4.2 K and show that we have the same high quality as for the photolithography junctions discussed in section 5.3. We did not observe any excess currents, so that the $I_c R_N$ products in the range of $1.75 - 1.93$ mV indeed indicate the strong Cooper pair tunneling as already mentioned above. R_{sg}/R_N ratios or V_m values were not evaluated, since they are usually determined at $T = 4.2$ K, so that no values for comparison are available. However, it was found that the R_{sg} values rose with sinking temperature, as can be expected according to the decrease in quasiparticle density according to (3.5). Detailed measurements of the subgap regime of these samples at $T = 20$ mK are discussed in sections 5.4.1 and 5.4.2.

The measurements at $T = 20$ mK also included $I_c(\Phi)$ modulations. In figure 5.10, the $I_c(\Phi)$ curve of sample B3 (for parameters see Table 5.2) is shown. The $I_c(\Phi)$ modulation could not be fitted with appropriate agreement by the Airy function (3.14). Instead, we found that the $I_c(\Phi)$ curve could be reproduced by Fourier transforming the integral critical current distribution $\Upsilon_c(x)$ shown in the inset. However, investigation with a scanning electron microscope (SEM) showed that the shape of the junction was indeed circular (a similar junction is shown in Figure 4.8b). Hence, the reason for this discrepancy is unclear. If the current flow through the junction was asymmetrically distributed, the $I_c(\Phi)$ modulation would be asymmetric as well and not go down to zero at all minima. Since this is not the case, a symmetric critical current density distribution and a high junction quality can be assumed.

A different trilayer with $j_c = 3$ kA/cm^2 led to junctions with higher I_c values. Although the critical currents in the range of $10 - 20$ µA were still somewhat suppressed at $T = 4.2$ K, a full characterization at this temperature was possible. The recorded gap voltages were all 2.8 mV or higher and the characteristic voltages $V_m = I_c R_{sg}$ were all larger than 50 mV. For Josephson junctions with areas > 1 µm^2, we observed values $V_m \geq 60$ mV, which can be attributed to a less suppressed I_c. The IV curve of a sample with an area of 0.5 µm^2 can be seen in Figure 5.11.

Since the R_{sg}/R_N ratios were very high for all critical current densities (see above) and high V_{gap}, $I_c R_N$ and V_m values were found at the exemplary critical current densities $j_c = 660$ A/cm^2 and $j_c = 3$ kA/cm^2, it can be concluded that a very high junction quality is obtained by the newly developed process.

5.4.1 Behavior Close to Retrapping

Using the four samples having a critical current density of $j_c = 660$ A/cm^2, we also investigated the behavior close to retrapping to the zero-voltage state. For this purpose, IV curves with a voltage bias were recorded at $T = 20$ mK. The measurement scheme used for the voltage bias setup can be seen in Figure 5.12. For all samples, two major current drops at voltages $V_{gap}/2 \approx 1.4$ mV and $V_{gap}/3 \approx 0.9$ mV could be seen (for example in Figures 3.4 and 5.14) and attributed to Andreev reflections (see section 3.1.3). The maximal subgap resistances $R_{sg,max}$ were obtained as illustrated in Figure 3.5 and are listed in Table 5.2.

Below $V_{gap}/3$, we were able to fit the data by (3.8) as shown by the blue curve in the inset of Figure 5.13. These fits contain the junction capacitance C, the line impedance Z_0, the theoretical critical current I_{c0} and the resistance R_{QP} seen by the DC quasiparticle current as parameters. The specific capacitance was recently determined by the measurement of

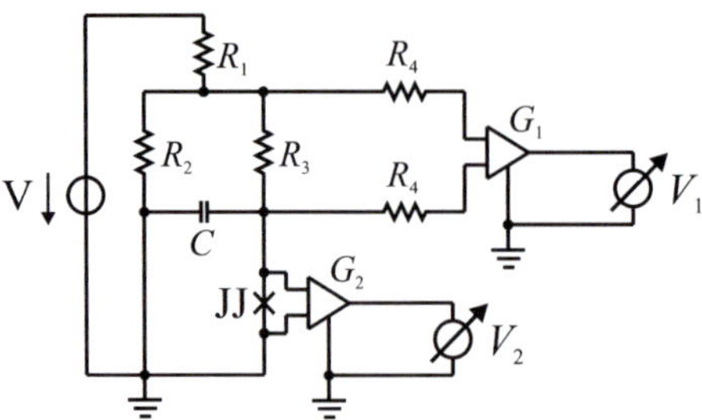

Figure 5.12: Setup used for the voltage bias measurements. $R_1 = 2.4$ kΩ, $R_2 = 51$ Ω, $R_3 = 510$ Ω, $R_4 = 1$ kΩ, $C = 330$ nF, $G_1 = G_2 = 500$. The voltage over the sample is given by $V_2/500$ while the current through the sample can be determined as $I = V_1/500/R_3$.

Fiske steps on a trilayer with an identical critical current density to be $c = 55$ fF/μm^2 [91]. The theoretical critical current I_{c0} was determined from the MQT measurements discussed in chapter 7, so that only R_{QP} and Z_0 remained as fit parameters.

The fitting results can be found in Table 5.2. The values for Z_0 correspond well to typical line impedances of 100 Ω. This shows that the employed RCSJ model describes our junctions accurately and that the different parts of the parallel circuit shown in Figure 2.1 are well-defined. The R_{QP} values that were determined are of the same order of magnitude, but systematically higher than the values for $R_{\mathrm{sg,max}}$. Since R_{QP} describes the dissipation a DC quasiparticle current in the voltage state and $R_{\mathrm{sg,max}}$ is supposed to give a measure for the damping in the zero-voltage state, it is not surprising that the values do not coincide.

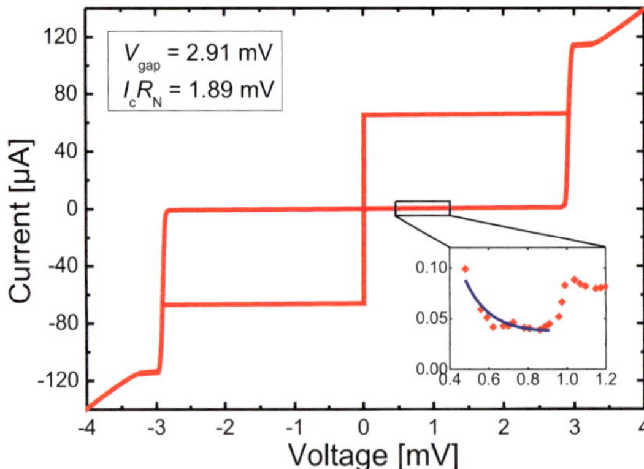

Figure 5.13: The *IV* curve of sample B3 shows the high quality regarding the $I_c R_{\mathrm{N}}$ ratio as well as low subgap currents. The inset shows a magnification of the subgap branch achieved by a voltage bias. Below the MAR step at $V_{\mathrm{gap}}/3$, the blue line is obtained by a fit with formula (3.8).

Table 5.2: Experimentally determined parameters for samples B1-4 at $T = 20$ mK. The theoretical critical currents I_{c0} were extracted from the MQT measurements described in chapter 7. R_{QP} and Z_0 were determined by fits using (3.8).

Sample name	Diameter [µm]	V_{gap} [mV]	I_{c0} [µA]	$I_c R_N$ [mV]	$R_{sg,max}$ [kΩ]	R_{QP} [kΩ]	Z_0 [Ω]
B1	1.9	2.88	19.1	1.75	54.0	133.6	86.4
B2	2.55	2.88	31.9	1.86	73.0	173.3	105.8
B3	3.6	2.92	68.1	1.93	21.1	44.0	88.1
B4	3.8	2.90	70.8	1.91	31.5	38.2	70.9

5.4.2 Estimation of Decoherence Rate Due to Quasiparticle Conductance

The same measurement run was used to determine the magnetic field dependence of the maximal subgap resistance $R_{sg,max}$. As explained in section 3.1.3, this should give a measure of the decoherence rate due to quasiparticle conductance in quantum devices fabricated with our newly developed fabrication process.

The measurement results can be seen in Figure 5.14. Figure 5.14b shows that the obtained $R_{sg,max}$ values increase systematically with a stronger suppression of the critical current. However, it can also be seen that we were not able to extend our investigation to the same regime of $I_c(\Phi) = 10^{-4} I_c(0)$ as in [64], so that values for the intrinsic quasiparticle resistance could not be determined. Instead, the measured $R_{sg,max}$ values seem to level off earlier than expected, which is due to the limited resolution of our voltage bias setup shown in Figure 5.12. In order to still obtain an estimate of the decoherence rate due to quasiparticle conductance, we systematically compared our data with the data presented in [64]. The excellent agreement in the $R_{sg,max}$ of $I_c(\Phi)$ dependence can be seen in Figure 5.14b. Furthermore, our $R_{sg,max}/R_N$ ratios at zero magnetic field lie between 1,160 and 2,980 for all investigated samples B1-4, which is in good agreement with values of 1,300 to 1,800

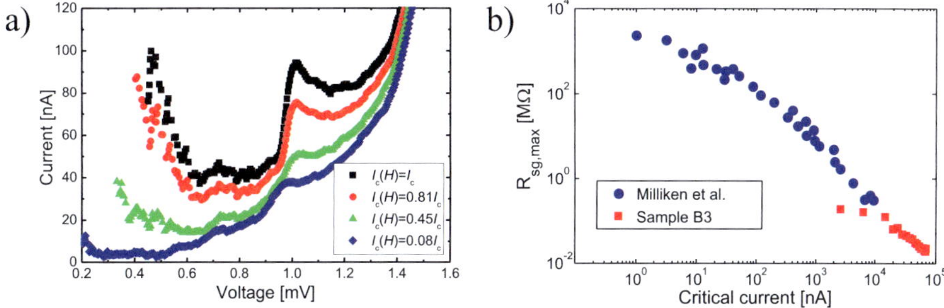

Figure 5.14: a) The subgap regime of sample B3 for different applied magnetic fields, measured with a voltage bias setup. b) The subgap leakage current decreases significantly for a more strongly suppressed critical current. Our data are in excellent agreement with the data presented in [64].

determined in [64] for samples from different foundries.

Hence, we assume that we can take over the conclusion reported in [64], where the decoherence rate from quasiparticle conductance $1/\tau_{\mathrm{qp}}$ was estimated using (3.9) to be as low as $1/\tau_{\mathrm{qp}} = 1/(100\ \mu\mathrm{s})$. Although this value is not directly determined from our measurements and only a rough estimate, we can certainly say that the coherence in our phase qubits (see chapter 9) should not be limited by quasiparticle conductance, since τ_{qp} is more than two orders of magnitude larger than the longest observed coherence times in phase qubits of around 600 ns [92].

5.4.3 Investigation of Scalability

Analogous to the investigations in section 5.3.1, chains of Josephson junctions were also fabricated with the process for sub-μm to μm-size junctions described in section 4.3. Due to the limited yield of SiO via lift-off discussed in section 4.3.2, a chain length of only 20 junctions was chosen. The *IV* curve of a chain containing 20 junctions each having a diameter of 2 μm can be seen in Figure 5.15. 19 junctions were recognized by the analysis algorithm, which corresponds to a yield of 95 %. Considering that not all junction chains could be measured due to the SiO lift-off yield of 92 % (see section 4.3.2), a total process yield of 87 % is obtained. This is certainly not sufficient for upscaling of the process for circuits containing thousands of junctions, but certainly enough for the quantum experiments performed in this thesis.

As can be seen in Figure 5.15, abnormal junctions were also observed for this process. This confirms the assumption that the occurrence of such junctions is not due to a limited precision of lithography, but are rather to the influence of plasma processes [82]. The quality parameters determined for the chain show that the high quality discussed above for single junctions is indeed characteristic for the newly developed fabrication process and observed throughout the chip.

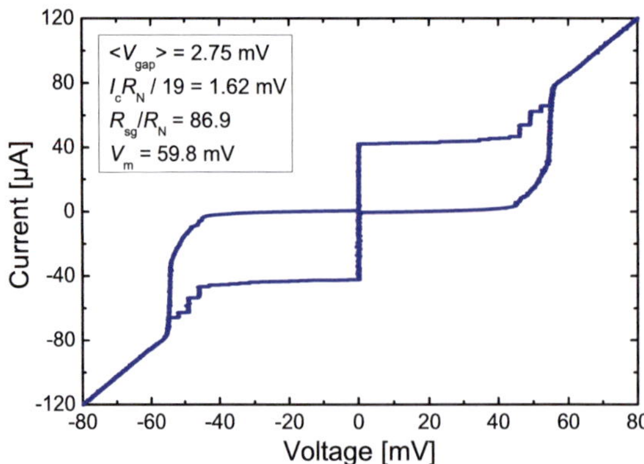

Figure 5.15: *IV* curve of a chain of 20 Josephson junctions with a diameter of 2 μm and a critical current density of $j_c = 1.5\,\mathrm{kA/cm^2}$, measured at 4.2 Kelvin.

5.4.4 Conclusions

In summary, it can be concluded that the sub-µm to µm-size Josephson junctions exhibit equally high quality as the junctions fabricated with the optimized photolithography process. The newly developed Al hard mask technique resulted in a precisely defined junction size and high-quality junctions down to areas of 0.5 µm^2. Detailed measurements of the subgap regime showed that the RCSJ model describes our junctions accurately and that all components influencing their impedance at high frequencies are well known. Furthermore, we extended these measurements by suppressing the critical currents with a magnetic field and found that the coherence times in qubits fabricated with the newly developed technology should not be limited by quasiparticle conductance. Finally, investigations of Josephson junction chains showed that scalability and yield should be high enough for all desired quantum experiments.

6 Measurement of Dielectric Losses in Amorphous Thin Films Used in Josephson Junction Fabrication

So far, the Josephson junction fabrication technology has been optimized with a focus on improving typical quality parameters deduced from DC characterization measurements. For the fabrication of superconducting quantum bits with long coherence times, however, the usage of low-loss dielectric layers is an additional requirement. Consequently, a measurement method for dielectric losses in thin films at GHz frequencies was developed within this thesis and the losses in different materials used for junction fabrication were measured. In this way, the losses in SiO and SiO_2 thin films (the materials available at the IMS) could be compared, but also further materials used in Josephson junction fabrication were investigated. This chapter starts by explaining the importance of low-loss dielectric layers for Josephson junction fabrication. Subsequently, the frequency dependence of dielectric losses is discussed. Then, a method is developed to reliably, quantitatively and quickly measure such dielectric losses in amorphous thin films. Details about the sample design and fabrication are given and finally, the measurement results are presented and discussed. Essential parts of this chapter have been published in [KSW$^+$10].

6.1 Introduction

It can be seen in Figure 6.1a that the insulation layer used in Josephson junction fabrication is placed between the lower electrode (M2a) and the wiring layer (M3). The junction electrodes are always larger than the junction itself (the additional overlap is called *idle region*, see Figure 6.1b), so that an extra capacitance in parallel to the actual Josephson junction capacitance is created. Consequently, the losses in the insulating layer will influence the device performance. This is even more true for structures with capacitively shunted junctions (such as superconducting phase qubits, see chapter 9), since most of the electric energy is stored in the capacitor and hence in the dielectric layer in this case.

Josephson junctions are employed for many applications involving microwave signals, so that low microwave losses in the structures are desirable in order to optimize the device performance. Such applications include Josephson voltage standards [3, 4], radiation sources [5], digital-to-analogue converters [7] or RSFQ circuits with passive transmission lines [93]. For superconducting qubits, low dielectric losses are even more critical, as high losses threaten the general operability of these devices. As has been shown in section 5.4.2, the coherence of superconducting phase qubits is not limited by quasiparticle conductance. Instead, Martinis *et al.* have pointed out recently that the microscopic origin of the dielectric losses are spurious dipoles or two-level systems (TLS), which couple to the qubit and

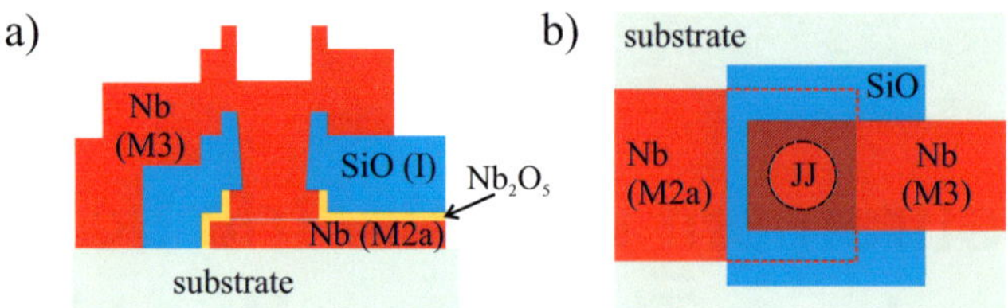

Figure 6.1: a) Schematic cross-section through a typical Josephson junction. The overlap of the bottom electrode M2a and the wiring layer M3 creates a plate capacitor like structure. b) The same junction in a schematic top view. The effective area of the mentioned plate capacitor is hatched and commonly known as *idle region*.

hence limit its coherence time [23, 94]. These TLS were found in dielectric thin films and the dependence of the losses on temperature and power was investigated [23, 94]. The dielectric losses on metal surfaces were also measured, but neither for amorphous thin films nor in a Josephson junction like geometry [95]. Furthermore, the processing of samples was found to influence the existing dielectric losses [96]. In none of those cases, the frequency dependence of the losses was measured.

Since we would like to build Josephson devices with low microwave losses in general and phase qubits with long coherence times in particular, we wanted to measure dielectric losses in thin films that can currently or potentially be used for Josephson junction fabrication. Since it is doubtful that the method employed in [23, 94] allows a reliable determination of the quantitative loss values, we decided to first develop a measurement method which allows a reliable, direct and fast determination of dielectric losses in thin films. Ideally, this method should also be able to extract both the losses in the volume of the thin film and on the metal/dielectric interfaces quantitatively. The development of such a measurement method is presented in section 6.3. It was used to investigate some materials typically used in Josephson junction fabrication at cryogenic temperatures and GHz frequencies. Furthermore, this newly developed method allows to measure the frequency dependence of dielectric losses, which yields information about the mutual coupling of the TLS, as will be shown in the following section.

6.2 The Frequency Dependence of Dielectric Losses

Within this thesis, the focus was put on comparing different thin films to each other and identifying the material with the lowest dielectric losses. For this purpose, measurements at $T = 4.2$ K are sufficient. In this regime, relaxation processes dominate the behavior of the TLS and are also responsible for the frequency dependence of the losses. At lower temperatures and low powers (i.e. in the qubit working regime), the behavior of the TLS is dominated by resonant absorption, which leads to entirely different properties. Such measurements have also been carried out on SiO thin films in the *Diplomarbeit* of Sebastian Skacel [Ska10] within the frame of this thesis. They revealed very interesting results about the properties of TLS, for example that their density of states is frequency dependent. However, in order to explain these results in detail, a circumstantial discussion of the theory of glass physics would be inevitable, so that for details the reader is referred to [Ska10] and [SKW$^+$11]. For the development of a technological process for macroscopic quantum experiments, which is the focus of this thesis, the results at $T = 4.2$ K are more important, since they allow to compare different dielectric materials used in Josephson junction fabrication and to optimize their deposition processes regarding minimal dielectric losses. Consequently, only phenomena related to relaxation processes are going to be discussed in this chapter.

Dielectric losses are measured as the loss tangent $\tan\delta = R/X$, which is the ratio of the real to the imaginary part of the complex impedance $Z = R + iX$. Consequently, the impedance can be written as

$$Z = X \cdot \tan\delta + i \cdot X \,. \tag{6.1}$$

Alternatively, dielectric losses are often given by the imaginary part χ'' of the complex electric susceptibility χ. The easiest model describing the frequency dependence of the losses is the Debye picture of non-interacting ideal dipoles with the same relaxation time τ, which leads to a frequency dependence [97]

$$\chi \propto \frac{1}{1 + i\omega\tau} \,. \tag{6.2}$$

In this picture, the losses are maximal at a peak frequency $\omega_{\mathrm{peak}} = 1/\tau$. Although some materials indeed show such a loss peak, pure Debye behavior is seldom observed in nature, and especially not for amorphous dielectrics [97]. Instead, it appears to be a remarkably common rule that the frequency dependence of dielectric loss for $\omega > \omega_{\mathrm{peak}}$ is given by fractional power laws [97–99]:

$$\chi'' \propto \omega^{n-1} \quad \text{with} \quad 0 \leq n \leq 1 \,. \tag{6.3}$$

This relation extends over several decades of frequency from the lowest audio and sub-audio range to the single-digit GHz range in a remarkably wide range of different materials [99] and is usually referred to as the 'universal law'. For higher frequencies, lattice mode excitations as well as electronic excitations will arise [99]. It can be assumed that (6.3) will describe our data for all investigated frequencies up to ~ 15 GHz, which is supported by our measurement results (see section 6.5). It has been shown that such a universal behavior is expected regardless of the electronically active species in the material, as long as low energy excitations are mainly responsible for the losses [99]. In [97, 99] the frequency dependence of the dielectric losses is discussed and it is shown that the exponent n of the power law depends on the coupling of the TLS to each other. For dipoles, which are expected to be present in amorphous dielectrics, the model predicts exponents of $n = 0$ for no interactions (Debye-like), $n \approx 0.5$ for nearest-neighbor interactions and $n > 0.6$ for many-body interactions [99].

6.3 Method of Measurement

The results of the $\tan\delta$ measurements should shed some light on the dielectric losses in applications of Josephson junctions. Consequently, the measurement method should in-

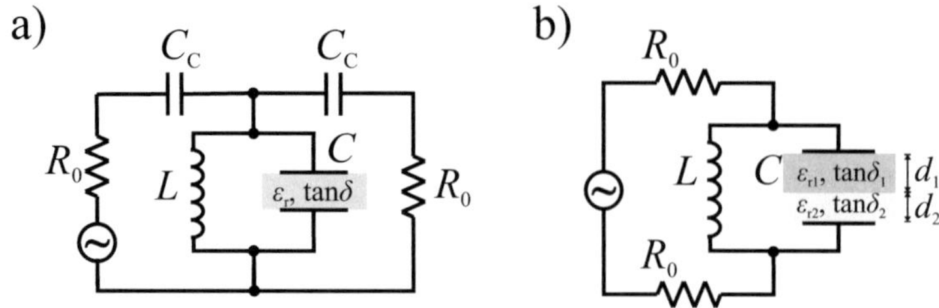

Figure 6.2: a) Schematics of the parallel lumped element *LC* circuit used for $\tan\delta$ measurements in [23, 94]. b) Schematics of the circuit used in this thesis. The pure *LC* circuit is observed in measurement and Q_0 can be directly read from the measured data.

volve the same materials (i.e. Nb electrodes), the same frequencies (i.e. in the GHz range) and the same temperatures (i.e. at 4.2 K or even below) that are used in device development and fabrication. The natural choice as tool of investigation are hence superconducting resonators.

When measuring the magnitude of transmission $|S_{12}|$ of such a superconducting resonator, one normally obtains the loaded quality factor Q_L, which is given by $1/Q_L = 1/Q_0 + 1/Q_C$. Here, Q_C is a measure of how much the resonator is decoupled from its environment. Q_0 is the intrinsic quality factor and denotes the loss in the resonator itself. It is composed of three contributions:

$$\frac{1}{Q_0} = \frac{1}{Q_\varepsilon} + \frac{1}{Q_{\mathrm{rad}}} + \frac{1}{Q_\rho}. \tag{6.4}$$

Here, $1/Q_\varepsilon = \tan\delta$ denote the dielectric, $1/Q_{\mathrm{rad}}$ the radiation and $1/Q_\rho$ the conductor losses. For a carefully designed superconducting resonator with matched lines and a closed housing, $1/Q_{\mathrm{rad}}$ is negligible. As the Nb surface impedance of films fabricated at the IMS accounts for $R_{\mathrm{s}} = 0.32\ \mu\Omega/\mathrm{GHz}^2 \cdot f^2$ [87], we can estimate $1/Q_\rho \approx 1 \cdot 10^{-7}$ at 100 MHz and $1/Q_\rho \approx 1.5 \cdot 10^{-5}$ at 15 GHz (all at 4.2 Kelvin). This is at least one order of magnitude lower than all measured $1/Q_0$ values (see Figure 6.5). Since R_{s} decreases exponentially with temperature [100] and $Q_\rho < 10^{-22}$ already at 300 mK, the described method would be especially suitable for the investigation of dielectric losses in the qubit working regime. The fact that conductor losses do not enter the measured Q_0 values has also been confirmed by simulations, as will be discussed in section 6.4. In the same section, it will also be shown that potential inductor losses do not enter the obtained Q_0 values. Altogether, we can safely assume that for this measurement method, (6.4) turns into $1/Q_0 = 1/Q_\varepsilon = \tan\delta$. That means that by measuring the transmission $|S_{12}|$, we can obtain the total dielectric losses at the resonance frequency f_{res} if we are able to extract the intrinsic quality factor from the measurement data.

Superconducting resonators have already been used for $\tan\delta$ measurements of thin films. Coplanar waveguides have been used with a dielectric layer on top [94], but in that case different materials contribute to the total losses, so that full knowledge of the losses in the substrate and a finite element analysis is needed in order to determine the dielectric losses in the thin film under investigation. As an alternative approach, lumped element *LC* circuits

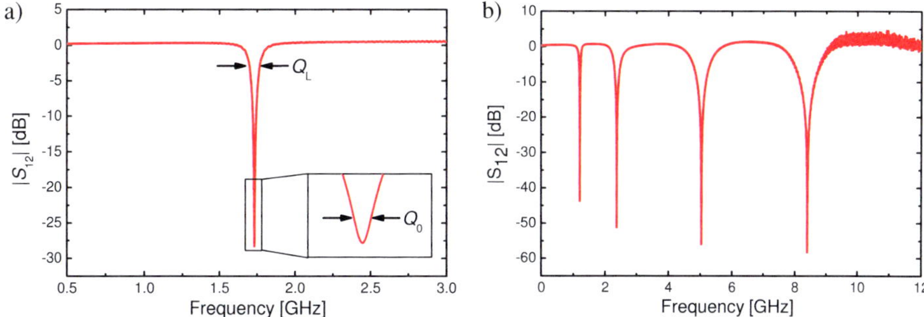

Figure 6.3: a) A typical measured resonance dip, here for a sample with Nb_2O_5 as dielectric, with illustration of Q_L and Q_0. b) Measurement of four resonator in series on the same chip, here with SiO as dielectric. Both curves were obtained at $T = 4.2$ K.

have been employed [23, 94]. In this case, the entire electric energy is stored between the capacitor plates where the material of interest is placed, so that substrate properties should not play a role. Additionally, the geometry of a plate capacitor is similar to the one in a Josephson junction and reproduces the electric field distribution found around a standard $Nb/AlO_x/Nb$ junction, so that the loss values determined in the resonators can be considered to be valid for Josephson circuitry (let aside the losses in the tunneling barrier itself). However, the circuit setup used in [23, 94] (see Figure 6.2a) requires knowledge of the coupling capacitors C_c and the load resistors R_0 in order to extract Q_0 from the measurement data. As we were looking for a method which allows to extract the $\tan\delta$ values directly from the measurement data without having to enter potentially faulty external parameters, we have developed a different method which allows a direct measurement of $\tan\delta$.

In our circuit setup (see Figure 6.2b), the *LC* circuit is directly connected to the load lines in series. This has the advantage that there is no parallel load resistance R_0, so that the intrinsic quality factor Q_0 can be directly extracted from the resonance dip as illustrated in Figure 6.3a. This is done by simply calculating

$$Q_0 = \frac{f_{\text{res}}}{\delta f}.$$

(6.5)

Here, δf is the bandwidth of the resonance dip at

$$|S_{12}| = \frac{\sqrt{2|S_{12,\text{min}}|^2}}{1 + |S_{12,\text{min}}|^2}$$

(6.6)

and $|S_{12,\text{min}}|$ is taken at the resonance [101]. It is obvious that besides a possible measurement error, no further error sources enter the determination of Q_0. Remembering that $1/Q_0 = \tan\delta$, we see that the dielectric losses in thin films can be directly and quantitatively measured with this new method. It should be mentioned that the usage of coupling capacitors is only necessary if high Q_L values are required. This is not the case here, so that we leave out such capacitors which only introduce complications to the Q_0 measurement as explained above.

Another big advantage of the method shown in Figure 6.2b comes from the fact that each resonator has full transmission outside of its resonance. This means that several resonators can be placed in series on the same chip. A transmission measurement for four resonators having resonance dips at different frequencies can be seen in Figure 6.3b. Such a setup is a paramount advantage for measurements in dilution refrigerators at low temperatures, since the cooldown and warmup times can be reduced by a factor of four. Even if one resonator had pinholes in the capacitor dielectric and hence did not work, it would just act as a shortcut so that all other resonators in the chain would still be measurable. Such a large bandwidth dilution fridge measurement setup was developed within the *Diplomarbeit* of Sebastian Skacel and is described in [Ska10] and [SKW$^+$11].

It has already been mentioned that TLS and hence dielectric losses can occur both in the bulk material as well as on metal surfaces or metal/dielectric interfaces. With the presented method, both contributions to the losses can be quantitatively measured by varying the dielectric film thickness and using the formula for losses in multi-layers (6.7). However, typical amorphous thin films used in Josephson junction fabrication have such high losses that the bulk losses will prevail and the interface losses will not be quantifiable.

6.4 Resonator Design, Sample Fabrication and Measurement Setup

More than 10 resonator geometries were designed in order to cover a wide frequency range from $\sim$150 MHz to $\sim$15 GHz. Even more resonance frequency values were obtained by varying the dielectric thickness and hence the capacitance values. Simulations of the designed structures were carried out with *Sonnet*[1]. Here, the dielectric losses and the permittivity of the substrate were systematically varied to confirm that no inductor losses enter the obtained Q_0 values. Indeed, we found that the relation $\tan\delta = 1/Q_0$ was valid for all employed resonator geometries, frequencies and substrate properties (with $\tan\delta$ being the losses in the dielectric between the capacitor plates). Also experimentally, some resonance frequencies were covered with different resonator geometries, meaning that the L and C values were varied relative to each other. We found that for all employed inductor designs (see below), the measured Q_0 value does not depend on the resonator geometry. This indicates that there was no redistribution of losses from the capacitor to the inductance, which would have been the case if inductor losses played a role.

The comparison between different resonator geometries was also employed to optimize the inductor design. It was found that meanders with too narrow stripes distorted the measurement results for Q_0, which can be attributed to capacitive coupling within the meander. Consequently, we only used rather wide meanders (where the line width and the spacing between the lines account for at least 50 μm) or straight lines as inductors. A typical resonator design can be seen in Figure 6.4a, in this case with a meander that is slightly too narrow, so that this sample was not employed for the actual measurements. However, this sample was used to investigate the power dependence of the resonance peak (see below), as it contains the narrowest lines and the sharpest bends.

---

[1]Sonnet Software Inc., 1020, Seventh North Street, Suite 210, Liverpool, NY 13088, USA

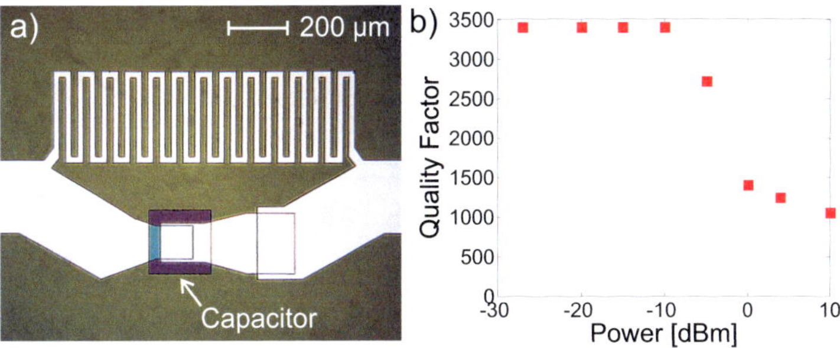

Figure 6.4: a) Photograph of an *LC* circuit with SiO as dielectric between the capacitor plates. This is the narrowest meander that was used to realize the inductance. b) Dependence of the measured quality factor on the applied power for the resonator shown on the left. The saturation is due to the high currents in the bends of the meander.

The resonators were realized on chips having a size of 9×9 mm^2, built into a metal housing and contacted to SMA connectors. Matching of the lines leading to the *LC* circuits is achieved by using a microstrip design. The inner line width of the microstrip lines depended on the substrate thickness and was adjusted so that an impedance of 50 Ω was reached for all samples. For the first measurement runs, brass housings were used with the bottom of the brass box acting as ground of the microstrip line. Later, some of the chips were measured again in copper housings, which lead to the observation of significantly higher Q_0 values. Obviously, the losses in the brass ground were higher than expected, meaning that strictly speaking, the relations $1/Q_\rho \ll 1/Q_\varepsilon$ and hence $1/Q_0 = \tan \delta$ do not hold. In the future, it is possible to ensure that indeed the absolute losses in the dielectric thin films are measured by sputtering Nb also on the backside of the chips in order to get a superconducting ground plane. For the measurements discussed in this chapter, however, the obtained $\tan \delta$ values were simply slightly overestimated. Since the goals of these measurements were to investigate the frequency dependence of the losses as well as to compare the losses in different materials and multi-layers with each other, the slight overestimation of the $\tan \delta$ values does not play a role. Measurements using copper housings confirmed that the frequency dependence of the losses did not depend on the housing material [Ska10]. Consequently, the measurements were completed using the brass housings and all results regarding frequency dependence, comparison of materials and losses in multi-layers can considered to be valid.

In addition to the materials used in our Josephson junction fabrication process, several other insulators have been investigated, which are either used by other groups in junction fabrication or might potentially be used for it. An overview of the sample fabrication is given in the following list:

- Resonators containing Nb$_2$O$_5$ as dielectric were fabricated at the IMS using the standard procedure described in section 4.2.3.
- Samples containing SiO were also fabricated at the IMS with the standard process described in section 4.3.2. The SiO thickness was varied between 200 and 400 nm and SiO was lifted-off in order to avoid any potential influence of reactive-ion-etching on the material properties.

- Samples containing SiO_2 were RF sputtered at the IMS in order to see whether this material would exhibit lower losses than SiO. For this purpose, the process gas mixture (containing Ar, O_2 and N_2) and the pressure during sputtering were varied systematically. The RF sputtering took place at a power of 100 W on water-cooled substrates and at rates of around 3 nm/min. Analogous to SiO, SiO_2 was also lifted-off. More details about these investigations can be found in the bachelor thesis of Wolfgang Heni [86].
- The samples containing silicon nitride (SiN_x) were fabricated at the Physikalisch-Technische Bundesanstalt (PTB) in Braunschweig as part of the standard junction process [81]. SiN_x was deposited by physically enhanced chemical vapor deposition (PECVD) at a temperature of 100 °C and a rate of 48 nm/min.
- Samples containing hydrogenated amorphous silicon (a-Si:H) were fabricated at the Istituto di Cibernetica "Eduardo Caianiello" (C.N.R.) in Naples, Italy. The dielectric films were deposited by PEVCD at $T = 250$ °C at a power of 5 W, a pressure of 200 mTorr and a rate of 18 nm/s. Here, the Nb base electrode as well as the a-Si:H layers were patterned by RIE. More details about the investigations on this material can be found in the *Diplomarbeit* of Sebastian Skacel [Ska10].

In all cases, the electrodes consisted of sputtered Niobium with a RRR of at least 3 and were clearly thicker than the London penetration depth in order to exclude any influence of the Nb properties on the measured quality factors.

After mounting the samples into the housings, the resonances were measured in the S_{12} parameter at 4.2 Kelvin with a network analyzer (Agilent E8361A) at the IMS. The power dependence of the quality factors was investigated in order to ensure that no overdrive occurred (see Figure 6.4b). Finally, all measurement data were obtained at powers below - 15 dBm. Thermal cycling as well as un- and remounting was carried out for several samples to make sure that the measured Q_0 values were independent of such external influences. As the Q_0 values could be reproduced to a precision better than 0.1%, our measurement procedure seems to be very robust.

6.5 Results and Discussion

6.5.1 Frequency Dependence of Losses in Single-Layers

For all investigated materials, no dependence of the losses on the dielectric film thickness was found. This is not surprising for such lossy amorphous materials and shows that the losses in the film volume exceed the interface losses by far.

The frequency dependences of $1/Q_0 \approx \tan\delta$ for single-layers of amorphous thin films of Nb_2O_5, SiO, SiO_2, SiN_x and a-Si:H are shown in Figure 6.5. The rather large spread of the measured $1/Q_0$ values for each material is apparent. This spread was not caused by the measurement procedure (see above) and found to be the same for samples from different or the same processing runs. It seems to be in the nature of such materials research and a large number of samples was measured to allow fitting to the data.

As illustrated in Figure 6.5 by the fits (solid lines), the universal law $\tan\delta \propto \omega^{n-1}$ is fulfilled for all investigated dielectrics with exponent values $0.57 \leq n \leq 0.68$. This is exactly

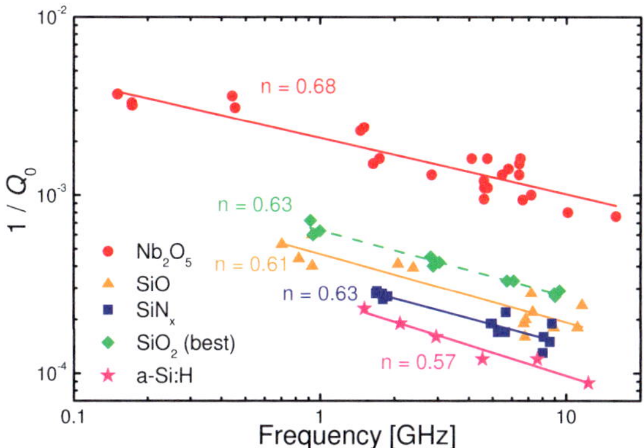

Figure 6.5: Frequency dependence of dielectric losses in different amorphous thin films usually used in Josephson junction fabrication, measured at $T = 4.2$ K. The data is in agreement with the universal law and the exponents n are in the expected range.

what is predicted within this model for dipoles with many-body interactions [99]. Such a behavior can be expected for the investigated amorphous and hence glass-like materials. Knowledge of the mutual coupling of the TLS would also be important for the understanding of decoherence mechanisms in superconducting qubits. In general TLS modeling [102] as well as qubit related publications [84, 103], the TLS are simply assumed to be non-interacting. At first sight, dielectric loss measurements in the qubit working regime should allow to reveal whether this is true or not. In this regime, however, the dominant loss mechanism will always be resonant absorption. In this case, only the TLS in resonance with the MW frequency will contribute to the losses so that the mutual coupling of the dipoles cannot be probed. Although it is not *a priori* clear whether the results obtained in the relaxation regime at $T = 4.2$ K can be directly transferred to the behavior of TLS in the qubit working regime, these results revealing the mutual coupling of the TLS should at least be considered in the context of qubit decoherence.

It can be seen in Figure 6.5 that the absolute values of the losses in different materials compare to each other as expected, for example from [23], where it was predicted that the losses in silicon compounds should decrease with increasing coordination number. We found that especially a-Si:H is a material with very low losses. Since a-Si:H and SiN$_x$ were not available at the IMS, it was not an option to use them in Josephson junction or qubit fabrication. Furthermore, the yield of working a-Si:H resonators was below 20 %, so that this material would not be useful for Josephson circuitry. More details about the investigations on a-Si:H can be found in the *Diplomarbeit* of Sebastian Skacel [Ska10].

As has been mentioned before, SiO$_2$ was investigated to find out if this material might exhibit lower losses than SiO and would hence be favorable for Josephson junction and especially qubit fabrication. But although a large parameter space was investigated, the dielectric losses were still found to be higher than for SiO, even for the optimal working point. The permittivity values ε_r, which were determined at the same time, indicate that the material obtained by RF sputtering is no ideal silicon dioxide but rather some SiO$_{2-x}$. More details about these investigations can be found in the bachelor thesis of Wolfgang

Heni [86]. After all, these results led to the decision that SiO remained in use in junction fabrication and was not replaced by SiO_2.

Figure 6.5 shows that Nb_2O_5 is a very lossy material. As it is a standard material in Nb based Josephson junction fabrication and helps to obtain low leakage currents, it would be desirable to still use it in device fabrication. For a constant anodization voltage of 40 V, the influence of the anodization time t_a on the dielectric losses in Nb_2O_5 was investigated. We found that the losses are practically constant and do not depend on t_a. It has been shown in section 4.2.3 that Nb_2O_5 tends to creep under the resist and can build a closed layer on top of small Josephson junctions and hence turn them defective. As this encroachment increases with t_a (see Figure 4.6b), short anodization times are favorable in the fabrication of small Josephson junctions. Consequently, it is an important result that short t_a times are not a disadvantage concerning dielectric losses. Furthermore, Nb_2O_5 might still be useful in the fabrication of low-loss Josephson devices despite its big losses, as will be discussed in section 6.5.2.

6.5.2 Losses in Multi-Layers

In Nb based Josephson junction fabrication, often two different insulating layers are used, namely Nb_2O_5, which grows out of the lower Nb electrode, and a sputtered or evaporated second material (in our case Nb_2O_5 is combined with SiO, see chapter 4). Consequently, it is important to know how the dielectric losses in such a multilayer are composed of the single material properties. To investigate this, we simply placed such a multi-layer between the capacitor plates of the LC resonator (see Figure 6.2), which is again equivalent to the configuration in a Josephson junction (see Figure 6.1).

As the two dielectrics are placed in series, their impedances must be added to $Z_{tot} = Z_1 + Z_2$. Knowing that the reactance of a capacitor is $X = 1/\omega C$ and employing (6.1), we find that the total losses are calculated by

$$\tan \delta_{tot} = \frac{\mathrm{Re}\,(Z_{tot})}{\mathrm{Im}\,(Z_{tot})} = \frac{d_1/\varepsilon_{r1} \cdot \tan \delta_1 + d_2/\varepsilon_{r2} \cdot \tan \delta_2}{d_1/\varepsilon_{r1} + d_2/\varepsilon_{r2}}. \tag{6.7}$$

Here, $\varepsilon_{r1,r2}$ are the permittivities and $d_{1,2}$ the thicknesses of each layer. This result is reasonable, as the individual $\tan \delta$ values are weighted by the electric energy stored in each dielectric $E = q^2/2C \propto d_i/\varepsilon_{ri}$ (here q is the stored electric charge). To validate this relation experimentally, we fabricated two sets of resonators ML1 and ML2 with different thicknesses of Nb_2O_5 and SiO. The measurement results were then compared to the theoretically expected values, obtained by using (6.7) and entering the fits for the single layers from Figure 6.5 as the $\tan \delta$ values. The very good agreement between measurement and theory can be seen in Figure 6.6. Also here, the measurements were carried out with brass housings, so that the $1/Q_0$ values are given. Analogous to the arguments discussed in the previous section, this should neither play a role for the frequency dependence of the losses, nor for the combination of losses in multi-layers.

This result has a direct implication for the usability of Nb_2O_5 in Josephson junction fabrication. Due to its high dielectric constant $\varepsilon_r \approx 33$, the lossy Nb_2O_5 might still be acceptable if used in thin layers and in combination with a material with low losses and a typical ε_r value (mostly $4 \lesssim \varepsilon_r \lesssim 8$).

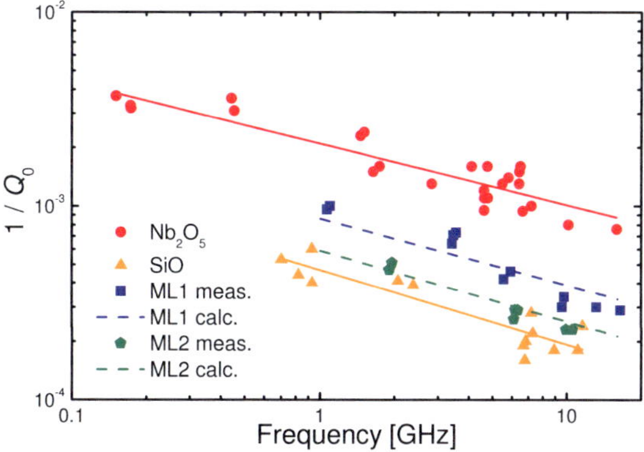

Figure 6.6: Dielectric losses in multi-layers of amorphous thin films. The dashed curves are the expected losses, calculated from the solid fits and equation (6.7). (ML1: 92 nm of Nb_2O_5 + 50 nm of SiO; ML2: 92 nm of Nb_2O_5 + 200 nm of SiO.)

6.6 Conclusions

In this chapter, the development of a reliable method for direct measurements of dielectric losses in thin films employing superconducting resonators was discussed. Our method does not require any fitting parameters and allows to measure the losses in the film volume as well as the metal/dielectric interfaces quantitatively. The losses in thin films of Nb_2O_5, SiO, SiO_2, SiN_x and a:Si-H have been measured at cryogenic temperatures and GHz frequencies, showing that the losses in the volume of such amorphous materials exceed the ones on the interfaces by far. Concerning the materials available at the IMS, it was found that thermally evaporated SiO exhibits lower losses than SiO_2, so that the former will be used for Josephson junction fabrication within this thesis.

For the first time, the frequency dependences of the losses under the above conditions were investigated and were found to be in good agreement with the universal law. The exponents obtained by fitting the data allow the conclusion that the dipoles exhibit many-body interactions, which is an important fact for their theoretical modeling. The investigation of the dielectric losses in multi-layers showed a good agreement between measurement and theoretical expectation. This allows the conclusion that even materials with high losses might still be used for Josephson junction fabrication in thin layers, as long as the have a very high permittivity (like Nb_2O_5). The measurement method introduced here allows to extend the measurements to further materials, temperatures and frequencies. The obtained results directly apply to many applications of Josephson junctions. In order to learn more about the behavior of the TLS in the qubit working regime (single-photon power, mK temperatures), corresponding measurements were carried out in the *Diplomarbeit* of Sebastian Skacel. As already discussed above, their results are not discussed here, but can be found in [Ska10] and [SKW⁺11].

7 Dependence of the MQT Rate on Junction Area and Magnetic Field

Macroscopic quantum tunneling (MQT) — often referred to as secondary quantum effect — is the manifestation of the quantum mechanical behavior of a single macroscopic degree of freedom in a complex quantum system. It was first observed in the 1980s [10, 33], but even today, it is still the standard experiment to investigate whether the fabricated junctions reach the quantum regime, and whether their quantum properties can be predicted correctly from the sample parameters. In this chapter, such measurements on Josephson junctions fabricated with the newly developed technological process (see section 4.3) are discussed. Furthermore, the dependence of the tunneling rate on junction size and magnetic field is investigated experimentally, which had never been reported before.

The chapter starts by giving a short introduction to the topic. Afterwards, the theoretical dependences of MQT on varying junction size, damping and magnetic field are discussed. Then, the procedure and setup of measurement are described in some detail. Afterwards, the investigated Josephson junctions are characterized carefully, and finally, the results of the MQT measurements are presented and discussed. Parts of this chapter have been published in [KSS11].

For MQT, the difference between the theoretical critical current I_{c0} and the switching current I_{sw} (see section 2.2.3) is crucial, so that this notation will be used here to emphasize the difference to the general critical current I_c.

7.1 Introduction

Since the phase difference over a Josephson junction φ is a macroscopic variable, circuits containing such junctions have been used as model systems for the investigation of quantum dynamics on a macroscopic scale. This research has recently led to the development of different types of superconducting quantum bits [12–16], which are promising candidates for the implementation of quantum computers. The starting point of this field was the observation of macroscopic quantum tunneling (MQT) in Josephson junctions in the 1980s [10, 33]. An introduction to this effect is given in section 2.2.3. MQT is the manifestation of the quantum mechanical behavior of a macroscopic degree of freedom, and is the main effect on which all quantum devices operated in the phase regime (such as phase qubits and flux qubits) are based. Consequently, the detailed understanding of MQT is not only interesting by itself, but also important for current research on superconducting qubits operated in the phase regime.

7.2 Theoretical Expectations

The basics of MQT have already been discussed in section 2.2.3. Now, more detailed theoretical considerations are carried out, in order to understand the expected behavior of MQT depending on junction size, damping and magnetic field.

7.2.1 Influence of Junction Size on MQT

For a Josephson junction, the critical current $I_c = j_c \cdot A$ and capacitance $C = c \cdot A$ are given by the critical current density j_c and the specific capacitance c of the trilayer and the area A of the junction. Both j_c and c are constant for a given trilayer.

By reformulating (2.19), we find that $\omega_{p0} = \sqrt{2\pi j_c/\Phi_0 c}$, meaning that for Josephson junctions on the same trilayer, the plasma frequency does not depend on their size. So at first sight, the crossover temperature (2.24) should also be independent of the junction size. In reality, however, the problem is more subtle, as one needs to take into account at which normalized bias current γ_{cr} the quantum tunneling rate (2.23) becomes significant. This is the point at which the plasma frequency (2.19) in (2.24) has to be evaluated. Since the height of the potential barrier $\Delta U \propto E_J \propto A$ is proportional to the Josephson junction size, a significant tunneling rate should be reached at different γ_{cr} values for junctions of different size. These points can be estimated by theoretically calculating (2.23) and converting it into a switching current histogram, as it would be observed in a real experiment. The probability distributions of switching currents $P(I)$ can be obtained from the quantum rate by equating [104]

$$P(I) = \Gamma_q \left(\frac{\mathrm{d}I}{\mathrm{d}t}\right)^{-1} \left(1 - \int_0^I P(u)\mathrm{d}u\right), \tag{7.1}$$

where $\frac{\mathrm{d}I}{\mathrm{d}t}$ is the ramp rate of the bias current. For the parameters of the junctions investigated

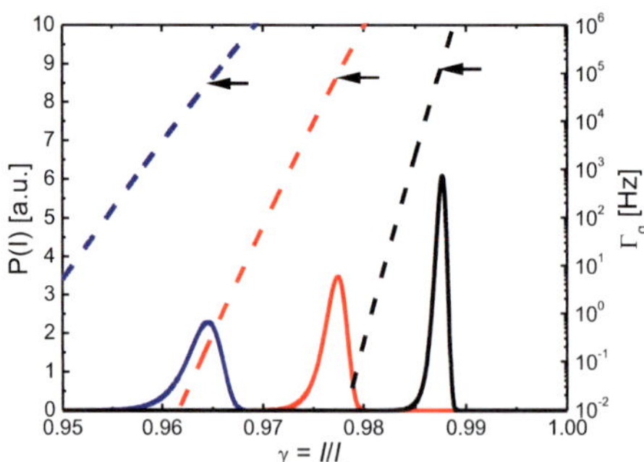

Figure 7.1: Theoretically calculated switching current distributions $P(I)$ (solid curves) and quantum tunneling rates Γ_q (dashed curves) for samples B1 (left) to B3 (right) having different diameters d (for parameters see Table 7.1). The difference in γ_{cr} (maximum position of $P(I)$), where quantum tunneling leads to escape from the potential well, is significant. The tunneling rates at these points (marked by arrows) are of the order of ≈ 100 kHz for all samples.

Table 7.1: Parameters of the investigated samples of type B and expected γ_{cr} and T_{cr} values. The critical current density accounts for $j_c \approx 660$ A/cm^2 while the specific capacitance of $c = 55$ fF/µm^2 was recently determined by the measurement of Fiske steps on a trilayer with identical critical current density [91]. Parasitic capacitances due to idle regions next to the junctions have been taken into account for all calculations in this chapter. For the calculation of T_{cr}, the actual measured critical currents were used.

Sample	Diameter [µm]	$\gamma_{cr}[]$	$T_{cr}[mK]$
B1	1.9	0.965	371
B2	2.55	0.977	323
B3	3.6	0.988	291
B4	3.8	0.988	277

in this chapter (see Table 7.1), the switching current distributions $P(I)$ were determined with a quality factor of $Q = 100$ and a current ramp rate of 100 Hz, according to our experiments (see below). They are shown in Figure 7.1, where it can be seen that the γ_{cr} values (the positions of the maxima of the distributions) significantly and systematically increase with the junction size. Evaluation of the $\Gamma_q(\gamma_{cr})$ values for the samples indicates that quantum tunneling will be experimentally observable at a rate of around $\Gamma_q \approx 10^5$ Hz.

Subsequently, the expected crossover temperature $T_{cr} = \hbar\omega_p(\gamma_{cr})/(2\pi k_B)$ was calculated. The sample parameters as well as the expected γ_{cr} and T_{cr} values are given in Table 7.1. It can be seen that the crossover temperature systematically decreases for increasing junction size. The change in T_{cr} is large enough to be observed experimentally. However, such a systematic study of the size-dependence of T_{cr} has never been carried out before.

7.2.2 Influence of Damping on MQT

The damping in a Josephson junction can be described by the dimension-free quality parameter

$$Q = \omega_p RC. \tag{7.2}$$

This quality parameter is often employed to describe the strength of the hysteresis in the current-voltage characteristics of a Josephson junction. In this case, one takes $Q = \omega_{p0}R_{sg}C$, with R_{sg} being the subgap resistance of the junction, so that $Q = \sqrt{\beta_C}$ with β_C defined in (2.21). Here, Q is size-independent, as $R_{sg} \propto 1/A$ and $C \propto A$. In the context of MQT, however, the dynamics take place at the plasma frequency ω_p, so that a complex impedance at that frequency $Z(\omega_p)$ has to be considered. For such an experiment as MQT, where the phase and not the charge is the well defined quantum variable, the admittance $Y(\omega_p)$ will be responsible for damping [105], so that the R in (7.2) will be given by $R = 1/\text{Re}(Y)$.

If the junction was an isolated system, the value of R in the context of MQT would be determined by the intrinsic damping in the zero-voltage state. The value which is typically taken as a measure for this is the maximal subgap resistance $R_{sg,max}$, which is simply the maximal resistance value which can be extracted from the nonlinear subgap branch (see section 3.1.3). In most experiments, however, the electromagnetic environment of the junction will be dominated by the lines leading to it. These leads can be assumed to have an impedance that is real and accounts for $Z_0 \approx 100$ Ω corresponding to typical transmission lines [33], but certainly not for more than the impedance of free space 377 Ω. As furthermore $Z_0 \ll R_{\text{sg,max}}$ and both contributions are in parallel (see Figure 7.2a), we can simply write $Q = \omega_{\text{p}} Z_0 C$ in this case.

Evidently, for small capacitance junctions (as in this experiment), the quality factor $Q = \omega_{\text{p}} Z_0 C$ will be limited to $Q \lesssim 10$ and additionally depend on the junction size like $C \propto A$. As we want to investigate the pure influence of the junction size on MQT, we would like to obtain very low damping as well as similar damping for all investigated junctions. In the implementation of phase qubits, current biased Josephson junctions have been inductively decoupled from their impedance environment by the use of circuits containing lumped element inductors and an additional filter junction [14]. In order to keep our circuits simple, we attempted to reach a similar decoupling by only using on-chip lumped element inductors right in front of the Josephson junctions (see Figure 7.2b). This setup leads to an admittance $Y = 1/R_{\text{sg,max}} + 1/(Z_0 + i\omega L)$, which yields

$$Y = \frac{R_{\text{sg,max}} Z_0 + Z_0{}^2 + \omega^2 L^2 - i\omega L R_{\text{sg,max}}}{R_{\text{sg,max}} (Z_0{}^2 + \omega^2 L^2)}. \tag{7.3}$$

As for (7.3), we find $\text{Re}(Y) \to 1/R_{\text{sg}}$ in the limit $\omega L \to \infty$, big enough lumped element inductances should decouple the Josephson junctions from the Z_0 environment and result in a high intrinsic quality factor $Q = \omega_{\text{p}} R_{\text{sg,max}} C$ even for switching experiments. Although it might be difficult to reach this limit in a real experiment, decoupling inductors should definitely help to increase the quality factor and move towards a junction size independent damping.

The damping in the Josephson junction influences the thermal escape rate (2.22) via the prefactor $a_t < 1$, which has been calculated for the first time by Kramers in 1940 [30]. In

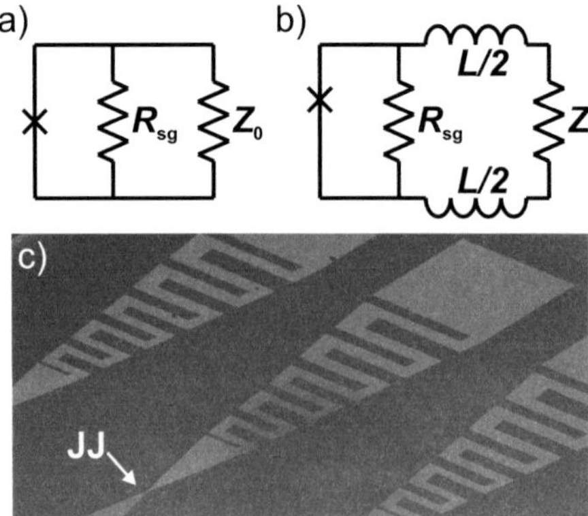

Figure 7.2: a) Typical impedance environment for switching experiments in a Josephson junction. As $Z_0 \ll R_{\text{sg}}$, the junction sees the impedance Z_0 at the plasma frequency. b) Lumped element inductors L can be used to decouple the junction from the line impedance Z_0, as discussed in the text. c) SEM micrograph of the electrode design used for the samples of type B.

the limiting case $Q \to 0$ (moderate to high damping), he found:

$$a_t = \alpha_{KMD} = \sqrt{1 + \left(\frac{1}{2Q}\right)^2} - \frac{1}{2Q}, \tag{7.4}$$

while in the opposite limit $Q \to \infty$ (very low damping limit), he found:

$$a_t = \alpha_{KLD} = \frac{36\Delta U}{5Q k_B T}. \tag{7.5}$$

More recently, Büttiker, Harris and Landauer [106] extended the very low damping limit to the regime of low to moderate damping finding the expression[1]

$$a_t = \frac{4}{(\sqrt{1 + 4/\alpha_{KLD}} + 1)^2}. \tag{7.6}$$

Additionally, damping reduces the crossover temperature according to [31, 107]

$$T_{cr,Q} = \frac{\hbar \omega_p}{2\pi k_B} \cdot \alpha_{KMD}. \tag{7.7}$$

A possible way to determine the quality factor Q for such quantum measurements is to extract it from spectroscopy data [33]. Unfortunately, for samples with such a high critical current density as used in the experiments described in this chapter, this turns out to be experimentally very hard. Hence, we will limit the analysis of the damping in our experiments to the MQT measurements. However, other groups have found a good agreement between the Q values determined by spectroscopy and by MQT [33, 108] and we hope to observe such a major increase of the quality factor due to the decoupling inductors that minor uncertainties in it should not play a role.

7.2.3 Influence of Magnetic Field on MQT

As discussed in section 3.2, the application of a magnetic flux Φ through a Josephson junction suppresses its critical current I_c. Since circular junctions are investigated in this experiment, the modulation will be that of an Airy curve shown in Figure 3.7b, where the critical current minimum is at $\approx 1.22\Phi_0$. The naive prediction for the crossover temperature would just consider the dependence $\omega_p \propto \sqrt{I_{c0}}$ and hence be

$$T_{cr,\Phi} = \sqrt{\frac{I_{c0}(\Phi)}{I_{c0}(0)}} \cdot T_{cr,\Phi=0}. \tag{7.8}$$

Here, $I_{c0}(\Phi)$ is given by (3.13) or (3.14), depending on the shape of the Josephson junction.

---

[1] Equation (7.6) does not describe the turnover from low damping to high damping. This turnover problem has been addressed by several authors (see for example [31] and references therein). In general, more precise expressions agree with (7.6) in the parameter regime of our samples within the experimental resolution.

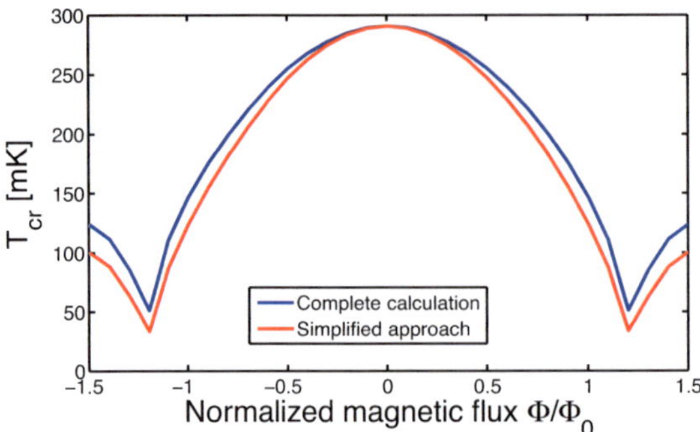

Figure 7.3: Theoretical expectation for T_{esc} as a function of magnetic flux for sample B3. The blue line was obtained by (7.8) and the red line by (7.9).

However, analogous to the considerations for the junction size, the change of the crossover point γ_{cr} has to be taken into account for a complete theoretical calculation. Consequently, the crossover temperature is given by the combination of (2.24) and (2.19) as

$$T_{cr,\Phi} = \frac{\sqrt{2e\hbar I_{c0}(\Phi)}}{2\pi k_B \sqrt{C}} \left(1 - \gamma_{cr}^2\right)^{1/4} , \tag{7.9}$$

where the crossover point γ_{cr} is determined by calculating $P(I)$ in the quantum regime via (7.1) as discussed before. Both equations have been plotted for sample B2 in Figure 7.3, where it can be seen that the difference is significant and should be observable experimentally. Furthermore, more complex physical phenomena like fluxons in the junctions or Fiske resonances might influence the observed T_{esc} dependence. However, no specific expectations were present before the experiments were started.

7.3 Setup and Procedure of Measurement

All samples were fabricated by the newly developed fabrication process for sub-µm to µm-size junctions discussed in section 4.3.

The employed measurement setup at the Institut für Festkörperphysik (IFP) can be seen in Figure 7.4. Special care was taken in design of the filtering stages in order to reach a low-noise measurement environment. In this context, especially the RC filters at the base temperature are crucial. The outside RF shield shown in in Figure 7.4 is essentially realized by the cryostat itself. In addition to this, the entire dilution refrigerator is placed in a screened room, only allowing electrical connections to the outside through metal tubes which are ten times longer than their diameter.

The goal of this measurement is to determine the escape rate Γ. In order to do so, we have measured the probability distribution $P(I)$ of switching currents. This was done by ramping up the bias current with a constant rate $\dot{I}$ and measuring the time t_{sw} between $I = 0$ and the switching to the voltage state with a Stanford Research 620 Counter, so that $I_{sw} = \dot{I} \cdot t_{sw}$ could be calculated. An Agilent 33250A waveform generator was used to create a sawtooth

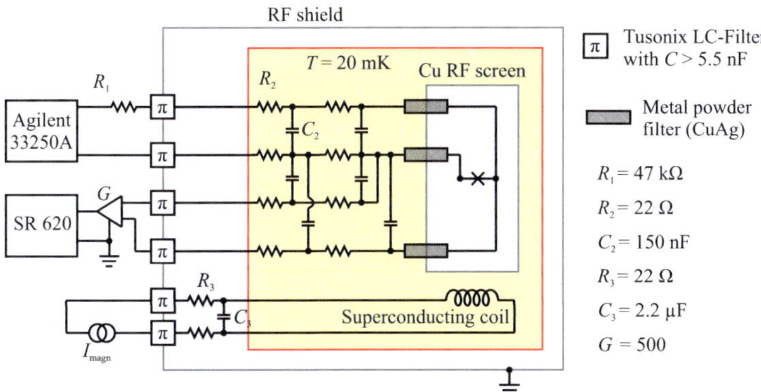

Figure 7.4: Schematic overview of the measurement system. The superconducting coil and the sample are inside a magnetic shield consisting of three nested cylindrical beakers, the middle one made from Pb, the two remaining ones from *Cryoperm*. Furthermore, the entire dilution refrigerator is placed inside a μ-metal shield at room temperature. The π-symbols denote commercial π-filters.

voltage signal, which was converted into the bias current by a resistor of 47 kΩ. Using this setup, I_{sw} could be measured repeatedly for each temperature. After doing so a large number of times, the switching current histograms $P(I)$ with a certain channel width ΔI were obtained as shown in the upper part of Figure 7.5. These histograms were then used to reconstruct the escape rate out of the potential well as a function of the bias current by employing [33, 104]

$$\Gamma(I) = \frac{i}{\Delta I} \ln \frac{\sum_{i \geq I} P(i)}{\sum_{i \geq I + \Delta I} P(i)} . \tag{7.10}$$

With Γ at hand, we can now determine the escape temperature T_{esc} by employing (2.22). In order to be able to rearrange this formula, we approximate the potential barrier in the limit $\gamma \to 1$ as $\Delta U = 4\sqrt{2}/3 \cdot E_J \cdot (1 - \gamma)^{3/2}$, so that we find

$$\left(\ln \frac{2\pi\Gamma(I)}{a_t(I)\omega_p(I)} \right)^{2/3} = \left(\frac{4\sqrt{2}E_J}{3k_B T_{esc}} \right)^{2/3} \frac{I_{c0} - I}{I_{c0}} . \tag{7.11}$$

Hence, by plotting the left side of (7.11) over the bias current I, we should obtain straight lines. Consequently, we can extract the theoretical critical current in the absence of any fluctuations I_{c0} as well as the escape temperature T_{esc} by applying a linear fit with slope ζ and offset Γ. We then find

$$I_{c0} = -\frac{\Gamma}{\zeta}, \tag{7.12}$$

$$T_{esc} = -\frac{4\sqrt{2}\Phi_0}{6\pi k_B \zeta \sqrt{\Gamma}} . \tag{7.13}$$

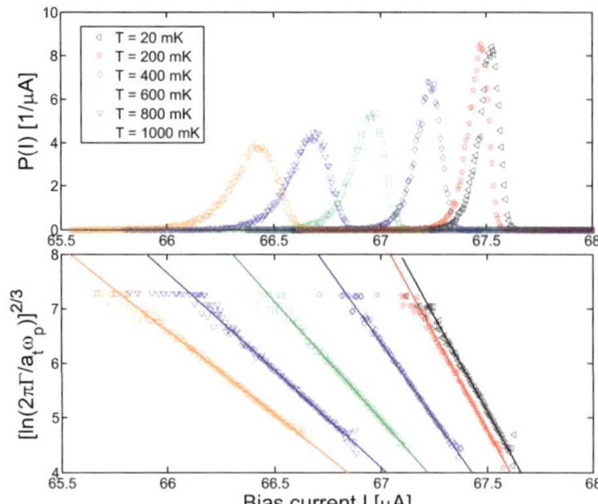

Figure 7.5: Top: The measured switching current histograms for sample B3 for selected temperatures. For increasing T, the switching currents decrease and the histograms broaden. Bottom: The plot obtained by applying (7.11) for the same sample. The fits allow to extract T_{esc} as well as I_{c0}.

Since I_{c0} enters (7.11) via E_J and ω_p, this fitting procedure has to be iteratively repeated until the value of I_{c0} converges. So strictly speaking, this procedure involves two fitting parameters, namely T_{esc} and I_{c0}. However, it turns out that I_{c0} is temperature independent within the expected experimental uncertainty (for all our measurements, the fit values of I_{c0} vary over the entire temperature range with a standard deviation of only around 0.09 %). Furthermore, the found I_{c0} values agree very well with the expected ones from the critical current density j_c of the trilayer and the junction geometry. Altogether, it can be said that the results for the main fitting parameter T_{esc} should be very reliable.

7.4 Sample Characterization

7.4.1 Sample A

The final size-dependent MQT measurements were performed on the samples of type B (see below). In order to characterize the sample quality, measurement setup and procedure, however, we first carried out a preliminary MQT measurement on sample A. This was a circular Josephson junction with a diameter of $d = 1.86$ µm and a critical current of $I_c \approx 12$ µA. As it was fabricated before the Al hard mask technique (see section 4.3.1) was developed, it was fabricated without anodic oxidation, so that SiO was the only insulating layer. The IV and $I_c(\Phi)$ curves can be seen in Figure 7.6 and show that the junction is of high quality. The side maxima of the $I_c(\Phi)$ modulation are lower than expected, but the entire curve is still symmetric. A similar effect has already been observed and discussed in section 5.4, where it was shown that a homogeneous tunneling barrier can still be expected for this junction.

In the IV characteristics, a strong current rise above $V \approx 1.5$ mV can be seen. Although the only difference between this junction and the usual high-quality samples presented in

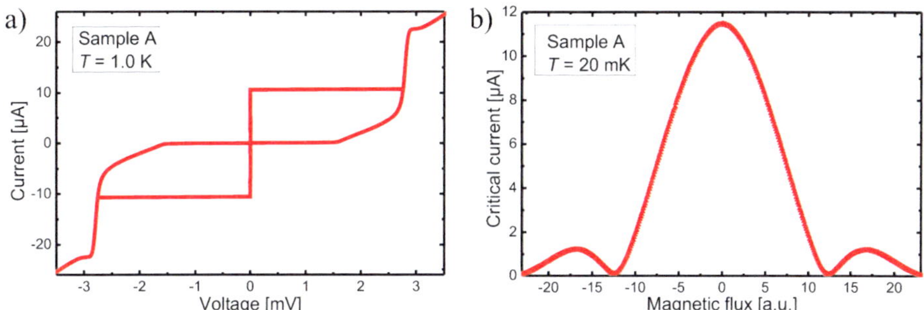

Figure 7.6: Characterization of sample A. a) The *IV* curve shows the high quality regarding the $I_c R_N$ ratio as well as low subgap currents. b) The $I_c(\Phi)$ measurement shows the expected behavior, indicating a homogeneous critical current distribution through the Josephson junction.

this thesis is the missing Nb_2O_5 insulation layer, this is no conclusive explanation for this effect. In any case, the subgap leakage currents close to $V = 0$, which should be important for MQT experiments, are still very low. Furthermore, the quality parameters $I_c R_N = 1.62$ mV and $V_{gap} = 2.76$ mV were determined from the *IV* curve and both indicate a high junction quality.

7.4.2 Samples B

The samples of type B were used for the size-dependent MQT measurements and implied a different electrode design containing lumped element inductors (see section 7.5.2). They were fabricated using the Al hard mask technique (see section 4.3.1), so that also the process

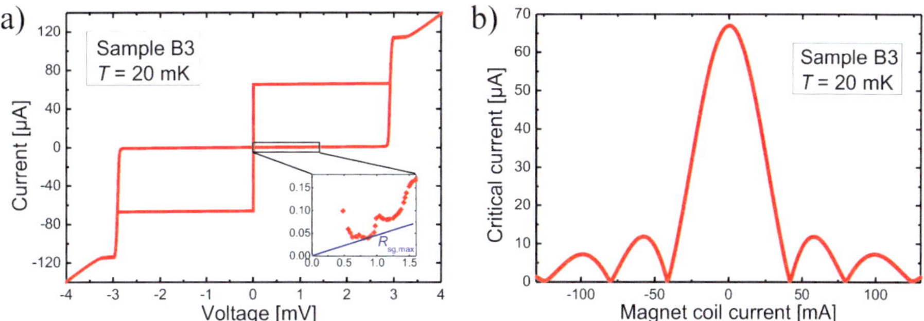

Figure 7.7: Characterization of sample B3. a) The *IV* curve shows the high quality regarding the $I_c R_N$ ratio as well as low subgap currents. The inset shows a magnification of the subgap branch achieved by a voltage bias. The blue line illustrates how the value for $R_{sg.max}$ was determined. The fact that the current rises for decreasing voltage at $V \approx 0.5$ V is due to the fact that the junction jumps back to a supercurrent $I \neq 0$ for $V = 0$ (see section 3.1.3). b) The $I_c(\Phi)$ measurement is discussed in more detail in section 5.4.

of anodic oxidation could be used. The junctions were circular in shape and their geometries are given in Table 7.1. In order to characterize the samples, $I_c(\Phi)$ curves, *IV* curves with a current bias as well as *IV* curves with a voltage bias were recorded. These samples were also taken to characterize the newly developed fabrication process for sub-µm to µm-size junctions, so that details about their characterization can be found in section 5.4 and their quality parameters are given in Table 5.2. Exemplary *IV* and $I_c(\Phi)$ curves for sample B3 can be seen in Figure 7.7. The values of the maximal subgap resistance $R_{sg,max}$ were extracted as illustrated by the blue line in Figure 7.7a. As already discussed in section 5.4, the quality parameters for all measured junctions indicate a very high quality.

7.5 Results and Discussion

7.5.1 Preliminary Measurement

In order to characterize our measurement setup, sample design and measurement procedure, we first measured one circular junction with a diameter of $d = 1.86$ µm and a critical current of $I_c \approx 12$ µA (sample A). The electrodes leading to the junction were simple wide lines and can be imagined as the envelope of the electrodes in Figure 7.2c. For the sawtooth bias current, a frequency of 10 Hz was employed and for each histogram, the switching current was measured 500 times. In the data analysis, we started with $a_t = 1$ and obtained a very good agreement with the expected behavior in the thermal regime. However, we still analyzed our data using a number of different Q values in (7.6) in order to see if we could determine the experimentally observed damping. This was done by calculating the deviation of T_{esc} from the bath temperature in the thermal regime:

$$\Delta T^2 = \sum_{T>500\,\text{mK}} (T_{esc} - T)^2 \tag{7.14}$$

and finding its minimum value regarding Q. For sample A, we found that (7.14) became minimal for $Q = 4$, so that the further analysis was carried out using this value. The calculated escape temperatures can be seen in Figure 7.8.

The theoretical calculation for this sample was carried out analogously to the calculation for the samples of type B described in section 7.2.1. A crossover temperature of 307 mK and a normalized switching current in the quantum regime of $\gamma_{cr} = 0.958$ are predicted. If the moderate damping of $Q = 4$ is taken into account, the crossover temperature according to (7.7) is reduced to $T_{cr,Q} = 0.883 \cdot T_{cr} = 271$ mK. Experimentally, a rather large spread of points was obtained, which can be attributed to the low number of measurements per temperature. We found $T_{cr} = 320$ mK as the average of the data points in the quantum regime, while the lower limit was at ≈ 280 mK. Furthermore, using the value of I_{c0} obtained from (7.12), we determined $I_{sw,cr}/I_{c0} = \gamma_{cr} = 0.957$. Despite the moderate measurement resolution, the values for T_{esc} and γ_{cr} are in good agreement with the predicted values.

Assuming we have $Q = 4$ as determined above, we can evaluate (7.2) and calculate an impedance of $R = 99.8$ Ω. This value is reasonable as it is clearly below the vacuum impedance of 377 Ω and corresponds to the expected value of $Z_0 \approx 100$ Ω for typical transmission lines. Furthermore, it is in excellent agreement with the values of Z_0 listed in Table 5.2. These were found by the analysis of the current rise close to retrapping to

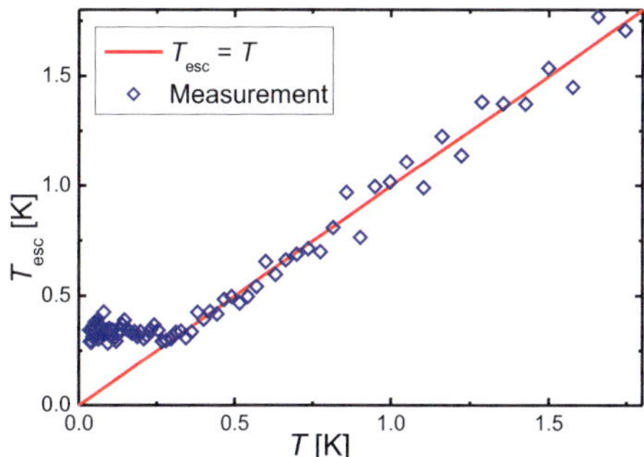

Figure 7.8: Preliminary MQT measurement on sample A. The crossover from the thermal to the quantum regime at $T_{cr} \approx 300$ mK is clearly visible.

the zero-voltage state, which is a clearly distinct physical experiment. Altogether, this means that the junction was in no way decoupled from the electromagnetic environment. In any case, this preliminary measurement made it clear to us that if we wanted to measure our small Josephson junctions in the low damping regime and obtain a high, possibly size-independent Q, we needed to change the electrode design and implement decoupling inductors according to (7.3).

7.5.2 Main Measurements: Damping

For our main measurements, we tried to learn our lessons from the preliminary investigation described above: The bandwidth of our filters was slightly increased, so that it was possible to apply a sawtooth signal with 100 Hz. This made it possible for us to increase the number of measurements for each histogram to 20,000 in order to get better statistics. Most importantly, however, we changed the on-chip electrode design for the samples of type B to the one shown in Figure 7.2c. The new design was based on the layout used for the lumped element inductors of the LC circuits with resonance frequencies in the GHz range investigated in chapter 6. Furthermore, simulations with *Sonnet*[2] confirmed that the meandered electrodes indeed act as lumped element inductors at the relevant frequencies $\omega_p(\gamma_{cr})$. The complex simulation with *Sonnet* gives an inductance of $L/2 \approx 1.65\,$nH (for one electrode) while the much simpler analysis with *FastHenry*[3] yields $L/2 \approx 1.8\,$nH. The Josephson junctions themselves were again circular in shape; an overview of the sample parameters is given in Tables 7.1 and 5.2.

For each sample, the experimentally observed quality factor Q was determined by analyzing the MQT data with various Q values in (7.6) and finding the minimum of (7.14). The corresponding quality factors were then used for the sample analysis. It can be seen in Figure 7.9 that the Q fitting values could be clearly identified. The evaluated Q values given in Table 7.2 show that we could drastically increase the quality factors with respect to the preliminary measurement. If we calculate the R values using (7.2), we find that they are clearly above the typical line impedance of $Z_0 \approx 100\ \Omega$ as well as the vacuum impedance

[2]Sonnet Software Inc., 1020, Seventh North Street, Suite 210, Liverpool, NY 13088, USA
[3]Fast Field Solvers, http://www.fastfieldsolvers.com

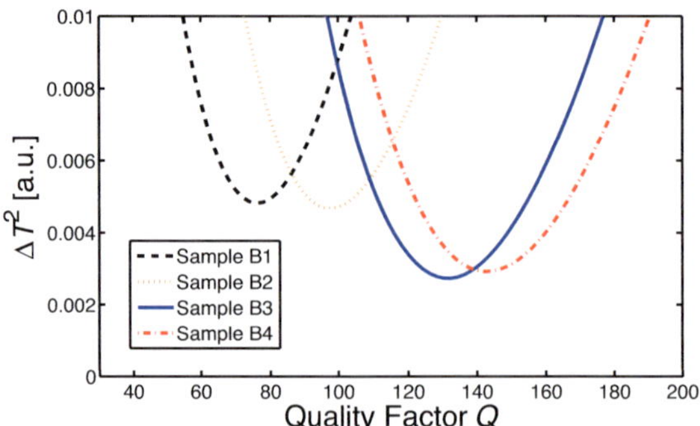

Figure 7.9: Determination of the experimentally observed quality factor Q for the samples of type B. The curves are minimal when the determined T_{esc} values deviate the least from the corresponding bath temperatures T in the thermal regime $T > 500$ mK.

of 377 Ω, which shows that we were indeed able to inductively decouple the junctions from their usual impedance environment. As expected, the determined R values are still clearly below the subgap resistances $R_{\mathrm{sg,max}}$, indicating that we have not reached the limit $\omega L \to \infty$. Instead, we are in the intermediate regime $Z_0 \ll R \ll R_{\mathrm{sg,max}}$, leading to the fact that Q still exhibits a slight dependence on the Josephson junction size (see Table 7.2). However, all junctions are in the low-damping regime, so that no influence of damping on the results should be present and experimental differences should indeed be due to the junction size. This can be seen by the fact that the correction in T_{cr} according to equation (7.7) is smaller than 1 % for all experimentally observed Q values. Altogether, we can state that we will be able to carry out our investigation of the size dependence of MQT with very low and nearly size-independent damping.

In addition to the rather qualitative considerations above, we performed a quantitative analysis employing formula (7.3). If we use $\omega = \omega_{\mathrm{p}}(\gamma_{\mathrm{cr}})$, take $R_{\mathrm{sg,max}}$ from Table 5.2 and assume that $Z_0 = 100\ \Omega$ (which was confirmed for our samples by different experiments, see the discussion in section 7.5.1), we can calculate the decoupling inductance L_{calc} for all samples. The values, given in Table 7.2, are a factor of around $2.3 - 2.5$ smaller than the simulation value of $L \approx 3.3$ nH, but of the right order of magnitude. For such a complex system, this is a surprisingly good agreement between simulation and theory on the one side and experimentally determined values on the other side. In summary, we can say

Table 7.2: The experimentally determined values characterizing the damping in the samples of type B. The L_{calc} values were determined from Q, the values given in Table 5.2 and formula (7.3). They are in good agreement with the design value of $L \approx 3.3$ nH.

Sample	Q [ ]	R [Ω]	L_{calc} [nH]
B1	76	1571	1.33
B2	98	1327	1.39
B3	132	1015	1.36
B4	143	1041	1.43

that we have successfully demonstrated that decoupling of the Josephson junction from its environment is also possible using only lumped element inductors.

7.5.3 Main Measurements: Crossover to the Quantum Regime

We now turn to the investigation of the crossover point from the thermal to the quantum regime and the influence of the junction size on it. As can be seen in Table 7.1, we expect a clear reduction of T_{cr} with increasing junction size. However, an experimental observation of lower crossover temperatures for smaller Josephson junctions having smaller critical currents could simply be due to current noise in our measurement setup. In order to exclude this, we artificially reduced the critical current of sample B3 by applying a magnetic field in parallel to the junction area. While unwanted noise should now lead to an increase in the observed T_{cr}, the physical expectation is a significantly reduced T_{cr} due to the lower plasma frequency. The result of this measurement can be seen in Figure 7.10 and Table 7.3. It is clear that we have a measurement setup exhibiting low noise, where the electronic temperature is indeed equal to the bath temperature, leading to an agreement between calculated and observed crossover temperature down to $T \approx 140$ mK, which was the lowest temperature we examined. This is clearly below any temperature needed for the comparison of the Josephson junctions of different sizes.

Finally, we measured the switching histograms for the four Josephson junctions of different sizes and evaluated the escape temperature T_{esc} as well as the theoretical critical current I_{c0}. This allowed us to determine the crossover temperature T_{cr} and the normalized crossover current $\gamma_{cr} = I_{sw,cr}/I_{c0}$. We indeed found a clear dependence of the crossover temperature on the junction size as can be seen in Figure 7.11. To compare the experimental γ_{cr} and T_{cr} values to the ones expected by theory, we now performed the theoretical calculation described in section 7.2.1 using the experimentally determined Q values and equation (7.7). All experimentally determined values are in excellent agreement with theory, as can be seen in Table 7.3.

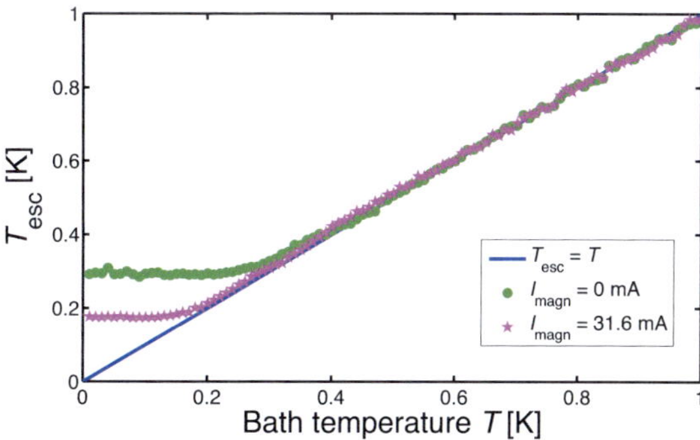

Figure 7.10: Escape temperatures for sample B3 with and without an applied magnetic field. For $I_{magn} = 31.6$ mA, which means that the critical current is suppressed to $I_c(\Phi) = 0.27I_c(0)$, a clear reduction of the observed crossover temperature T_{cr} is observed.

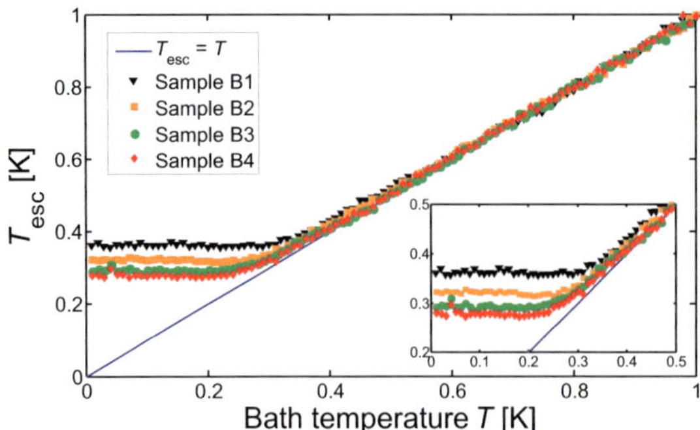

Figure 7.11: Escape temperatures for all samples of type B. The crossover to the quantum regime is very clear in each measurement. The inset shows a magnification of the quantum regime. The reduction of the crossover temperature T_{cr} with increasing JJ size is clearly visible.

Table 7.3: The experimentally determined values characterizing the crossover from the thermal to the quantum regime in comparison with the theoretical expectations for all measurements.

Sample name	I_{c0} [μA]	$I_c(\Phi)/I_c(0)$ [ ]	$\gamma_{cr,theo}$ [ ]	$\gamma_{cr,exp}$ [ ]	$T_{cr,Q,theo}$ [mK]	$T_{cr,exp}$ [mK]
B1	19.1	1	0.965	0.970	368	362
B2	31.9	1	0.977	0.981	322	321
B3	68.1	1	0.988	0.990	290	294
B3	35.5	0.52	0.984	0.987	223	236
B3	18.5	0.27	0.979	0.983	172	176
B3	9.14	0.13	0.973	0.975	129	147
B4	70.8	1	0.988	0.990	276	278

7.5.4 Main Measurements: Magnetic Field Dependence

For all investigated samples, the escape temperature was also measured as a function of applied magnetic flux through the junctions in the quantum regime at $T = 20$ mK. Such a systematic study has never been reported before. The theoretical expectation in this case is discussed in section 7.2.3.

The experimentally determined escape temperature T_{esc} as a function of magnetic flux for samples B2 and B3 can be seen in Figure 7.12. In order to plot the data over the normalized magnetic flux, it was assumed that the first minimum of the $I_c(\Phi)$ modulation lies at $1.22\Phi_0$, as should be the case for circular junctions. First, it should be noted that the data are absolutely symmetric and very clean. This is remarkable for such a sensitive quantum experiment. Furthermore, the agreement between the data and the theoretical expectation according to (7.9) is excellent up to about one flux quantum, while (7.8) does not describe the experimentally observed behavior sufficiently (the measured critical currents were the only experimental parameter entering the calculations). As can be seen in Figure 7.13, the

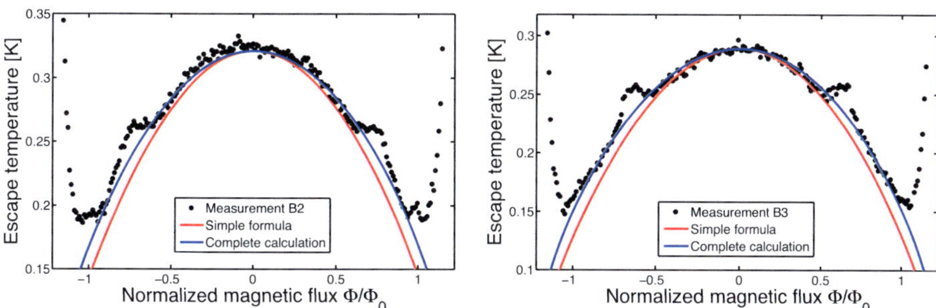

Figure 7.12: Escape temperature as a function of magnetic flux for sample B2 (left) and sample B3 (right). The red lines are theoretical calculations with (7.8), while the blue lines were obtained using (7.9).

significant variation of γ_{cr} with applied flux requires the use of (7.9). For magnetic flux values beyond $\pm\Phi_0$, the determined T_{esc} values diverge. At this point, the switching current histograms exhibited a second peak, which was situated at currents slightly smaller than for the main peak. This effect might either be due to the fact that unknown processes take place, for example by the formation of flux quantum vortices in the junction, or simply due to the very small critical currents and hence a stronger influence of noise on the measurement.

Additionally to the expected behavior given by (7.9), for each sample two peaks in T_{esc} are visible at $\approx \pm0.7\Phi_0$. At these flux values, no second switching peaks were observable in the switching current histograms, so that it can be assumed that the physical process behind this effect is different from the one leading to the divergence at $\Phi \approx \pm\Phi_0$. Since the T_{esc} values between $\approx 0.7\Phi_0$ and $\Phi \approx \Phi_0$ are in excellent agreement with (7.9) again, it can be excluded that the observed peak structures are due to noise, disturbances or any other problems with the measurement system.

The search for a theoretical explanation of this effect is currently still ongoing. The data were compared to a recently presented theory by Ovchinnikov *et al.* [109, 110], but were

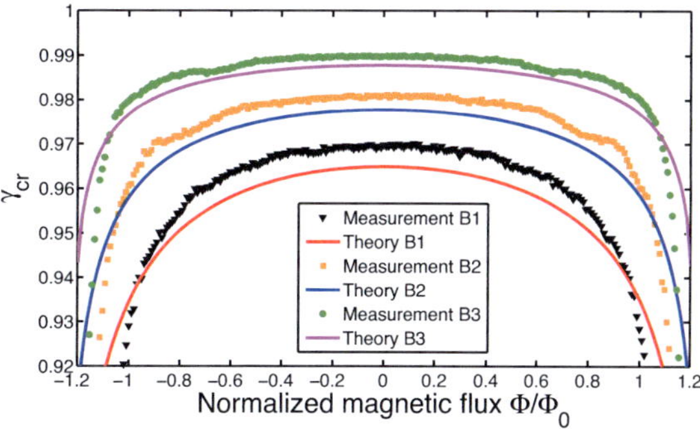

Figure 7.13: Theoretical and experimental values of the magnetic flux dependent normalized crossover current γ_{cr} for samples B1, B2 and B3.

not found to be in agreement with it [111]. Recently, a different explanation was suggested by A. V. Ustinov: The rise in T_{esc} might be due to the occurrence of Fiske resonances [25], which is a form of self-resonance in the junction. While the quantum fluctuations leading to the escape from the potential well can be considered to have a white spectral distribution, the occurrence of a self-resonance in the junction would lead to an additional fluctuation peak at a distinct frequency. Consequently, the rise in fluctuation amplitude would lead to the observation of a higher escape temperature. Since the first Fiske resonance is located at $\Phi = 0.7\Phi_0$ [25], this working hypothesis is in excellent agreement with the data and will be investigated in more detail in the future.

7.6 Conclusions

In this chapter, systematic macroscopic quantum tunneling (MQT) experiments for varying junction area and magnetic field have been discussed. In order to allow optimal comparability, the investigated samples were fabricated on the same chip. Thorough characterization before the actual quantum measurements revealed that the junctions exhibit a very high quality. It was shown that we could significantly decrease the damping at frequencies relevant for MQT by the use of lumped element inductors, which allowed us to perform our study in the low damping limit. The crossover from the thermal to the quantum regime was found to have a clear and systematic dependence on junction size, which is in perfect agreement with theory. In addition, the dependence of the escape temperature with applied magnetic flux agreed excellently with theory as well. Unexpected peak structures in T_{esc} at $\Phi \approx \pm 0.7\Phi_0$ might be due to the occurrence of Fiske resonances in the junctions. All these experiments confirm that the Josephson junctions fabricated with the newly developed fabrication process are excellently suited for quantum experiments.

8 Dynamics of LC Shunted Josephson Junctions

It has already been seen in the previous chapter that Josephson junctions are sometimes shunted by inductances in order to control their behavior. As will be shown in chapter 9, capacitive shunting is also a common technique in Josephson quantum circuitry. Furthermore, Josephson junctions in high-temperature superconductors often have stray capacitances and inductances which cannot be avoided in fabrication. Altogether, it can be said that the understanding of the quantum dynamics of Josephson junctions shunted by capacitors and inductors is important for many applications. Furthermore, such a system is much larger than for example a superconducting qubit, so that it is interesting to see whether it still behaves as one single quantum system. Consequently, the quantum dynamics of such LC shunted Josephson junctions (LCJJs) are investigated in this chapter. After a short introduction, the recently developed theory of such a system is discussed. Then, details of the sample design and simulation are given. In the following, the sample fabrication and pre-characterization are reported. After some short remarks about the measurement setup and procedure, the results are presented and discussed. Parts of this chapter have been published in [KBLS11].

8.1 Introduction

Recently, macroscopic quantum tunneling (MQT) was observed in d-wave $YBa_2Cu_3O_{7-\delta}$ (YBCO) Josephson junctions [112], which was the first that this effect could be measured in a high-T_c device. The YBCO Josephson junctions were fabricated on a (110) $SrTiO_3$ (STO) substrate. It was found that the behavior of the YBCO junction could not be described by the classic RCSJ model (see section 2.2.2). Instead, the stray capacitance C_s of the electrodes, due to the large dielectric constant $\varepsilon_r > 10,000$ of the STO substrate at low temperature, and the stray inductance L_s of the electrodes had to be taken into account by a modified RCSJ model [113, 114]. This modified model leads to a two-dimensional potential for the Josephson particle with two energy scales for the quantum levels in the potential well. Classical resonant activation in such a Josephson junction embedded in an LC circuit was also investigated theoretically [115]. Although the modified model was able to describe the quantum behavior of the YBCO junction, only the lower plasma mode could be observed [113, 114].

The implementation of high-temperature Josephson junctions in classical and quantum electronics requires a proper understanding of the influence of these stray elements on the device dynamics. For example, high-T_c junctions have been suggested for the realization of high performance quantum bits taking advantage of the intrinsic properties of the d-wave order parameter symmetry (known as 'quiet qubits') [116]. But a controlled design of the parameters L_s and C_s, which would be necessary for a systematic experimental study of their influence on the quantum dynamics, is not possible for high-T_c Josephson junctions. Instead, such an LCJJ system can be created artificially using low-T_c Josephson junctions, for which a very precise parameter control is available.

Furthermore, in low-T_c quantum circuits such shell circuits containing large capacitors and/or large inductors are intentionally used to influence the properties of the devices. For example, large shunting capacitors are used to reduce the plasma frequency of phase qubits [117] or readout SQUIDs [118]. Other designs use an inductor and capacitor as an isolation network to protect the qubit from the low impedance environment causing dissipation [14]. Also in this thesis, decoupling inductors were used for MQT (see section 7.2.2), and a shunting capacitor was used to adjust the energy levels of a superconducting phase qubit (see section 9.2.1). Altogether, it can be said that the understanding of the quantum dynamics in an LC shunted Josephson junction is also of paramount importance for many low-T_c quantum devices.

The goal of the experiment described in this chapter is to systematically study the influence of this kind of shell circuit on the dynamics of a Josephson junction in the quantum limit. Here, the good parameter control of the fabrication process discussed in chapter 4 is a crucial point for parameter variation and comparison between theory and experiment. In detail, it will be interesting to see whether indeed both energy scales of the two-dimensional potential are observable, since this has never been shown before. Furthermore, it will be investigated if such a large system (which is much larger than the YBCO junction in the original experiment) will behave as one single quantum system and not show the behavior of the single junction anymore.

8.2 Theory

The circuit schematics of such an LC shunted Josephson junction with critical current I_c are shown in Figure 8.1a. It can be seen that the normalized bias current $\gamma = I/I_c$ is divided into two parallel contributions: one branch containing the shell capacitor C_s and the other branch with the shell inductor L_s and the Josephson junction in series. Over each element in the circuit, a superconducting phase difference can be related to the current flowing through it. This yields $I_J = I_c \sin \varphi_J$ for the Josephson junction, $I_{C_J} = C_J \frac{\Phi_0}{2\pi} \ddot{\varphi}_J$ for the junction capacitance, $I_{C_s} = C_s \frac{\Phi_0}{2\pi} \ddot{\varphi}_s$ for the shell capacitance and $\varphi_L = \frac{2\pi}{\Phi_0} L_s I_s$ for the shell inductance. Employing Kirchhoff's circuit laws and the relation $\varphi_s = \varphi_L + \varphi_J$, the variable φ_L can be eliminated and the equations of motion of the circuit are obtained [114]:

$$\ddot{\varphi}_J + \sin \varphi_J + \frac{\varphi_J - \varphi_s}{\beta} = 0, \tag{8.1}$$

$$\vartheta^{-1} \ddot{\varphi}_s + \frac{\varphi_s - \varphi_J}{\beta} = \gamma. \tag{8.2}$$

Here, $\vartheta = C_J/C_s$, $\beta = L_s/L_{J0}$ and $L_{J0} = \hbar/(2eI_c)$ is the prefactor of the bias current dependent Josephson inductance L_J defined in (2.12). It can be seen that in contrast to an unshunted Josephson junction, two phase variables describe the dynamics of the circuit. This means that instead of the one-dimensional washboard potential shown in Figure 2.2,

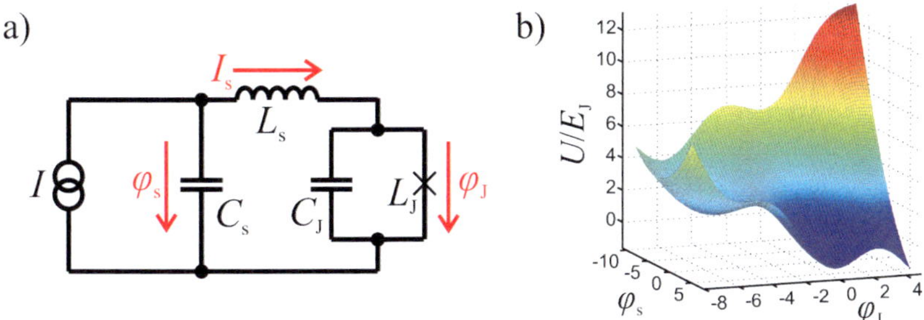

Figure 8.1: a) Circuit schematics of an *LC* shunted Josephson junction. b) Two-dimensional potential for the system shown in a) for $\vartheta = 0.1$, $\beta = 10$ and $\gamma = 0.5$.

the phase moves in a two-dimensional potential landscape. Such a potential is shown in Figure 8.1b for $\vartheta = 0.1$, $\beta = 10$ and $\gamma = 0.5$.

In the one-dimensional washboard potential of a single junction, the energy scales with the bias current dependent plasma frequency $\omega_\mathrm{p} = 1/\sqrt{L_\mathrm{J}C_\mathrm{J}}$. This energy scale can be explained by the formation of standing waves according to the Schrödinger equation. In a two-dimensional potential as for the LCJJ system, standing waves can form in two orthogonal directions, so that two energy scales will be present. These can be derived from (8.1) and (8.2) and were found to be [114]

$$\omega_\pm = \omega_{p0} \sqrt{\frac{1 + \beta\sqrt{1-\gamma^2} + \vartheta \pm \sqrt{(1 + \beta\sqrt{1-\gamma^2} + \vartheta)^2 - 4\beta\vartheta\sqrt{1-\gamma^2}}}{2\beta}}, \qquad (8.3)$$

with $\omega_{p0} = 1/\sqrt{L_{J0}C_\mathrm{J}}$. Since expression (8.3) is rather complex, it is helpful to evaluate it in different limits. The most trivial case is $\vartheta \gg 1$ and $\beta \ll 1$, where the shell capacitance and inductance are much smaller than the intrinsic values of the Josephson junction. Here, the lower mode is simply given by the junction's plasma frequency $\omega_- = 1/\sqrt{L_\mathrm{J}C_\mathrm{J}}$ and the upper mode is given by the resonance frequency of the shell circuit $\omega_+ = 1/\sqrt{L_\mathrm{s}C_\mathrm{s}}$. This reflects the fact that the modes of the junction and the *LC* circuit are decoupled from each other in this limit.

Physically more interesting is the limit where the shell capacitance is much larger than C_J, i.e. $\vartheta \ll 1$. This case is also present in many quantum devices containing Josephson junctions (an example is the phase qubit investigated in this thesis, see section 9.2.1), so that an the understanding of the dynamics in such a system is very important. In this case, the upper and the lower mode can be approximated as [114]

$$\omega_- = \omega_{p0} \sqrt{\frac{\vartheta}{\frac{1}{\sqrt{1-\gamma^2}} + \beta}} = \frac{1}{\sqrt{(L_\mathrm{s} + L_\mathrm{J})C_\mathrm{s}}}, \qquad (8.4)$$

$$\omega_+ = \omega_{p0} \sqrt{\frac{1}{\beta} + \sqrt{1-\gamma^2}} = \sqrt{\left(\frac{1}{L_\mathrm{J}} + \frac{1}{L_\mathrm{s}}\right)\frac{1}{C_\mathrm{J}}}. \qquad (8.5)$$

It can be seen that the lower mode ω_- is equivalent to a serial circuit of the elements L_J, L_s and C_s while the upper mode is equivalent to a parallel circuit of L_J, L_s and C_J. It

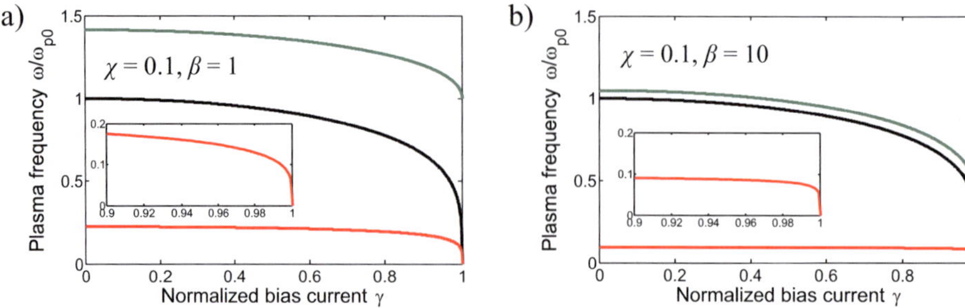

Figure 8.2: Plasma frequency ω_p for a single junction (black), upper normal mode ω_+ of the LCJJ system (green) and lower normal mode ω_- of the LCJJ system (red). The inset show a magnification of the lower mode close to $\gamma = 1$. a) For $\vartheta = 0.1$ and $\beta = 1$. b) For $\vartheta = 0.1$ and $\beta = 10$.

follows directly that for $\beta \gg 1$ (i.e. $L_s \gg L_J$) the modes will again be decoupled and the eigenfrequencies of the Josephson junction $\omega_+ = 1/\sqrt{L_J C_J}$ and of the shell circuit $\omega_- = 1/\sqrt{L_s C_s}$ will be obtained. So the most interesting case is the one for $\vartheta \ll 1$ and not too large values of β.

For the two cases $\beta = 1$ and $\beta = 10$, the theoretically expected eigenfrequencies ω_- and ω_+ are shown in comparison with the pure Josephson plasma frequency ω_p in Figure 8.2. It can be seen that the eigenfrequencies of the LCJJ system should be clearly distinguishable from that of a single junction, so that a clear criterion is available to evaluate whether the entire circuit acts as one quantum system in spectroscopy experiments or whether the single components are still observed. Furthermore, it is already visible that ω_- and ω_+ approach the eigenfrequencies of the Josephson junction and the shell circuit for $\beta = 10$. Since the shell circuit resonance frequency $1/\sqrt{L_s C_s}$ is not bias current dependent, this leads to a significantly changed curvature of the mode ω_-, as is shown in the insets of Figure 8.2.

For spectroscopy experiments, it is likely that processes of higher order will also be observed. This can be multi-photon and/or multi-level processes as described in section 2.2.4. According to (2.25), this leads to transitions at microwave frequencies $\omega_{MW} = \frac{p}{q}\omega_+$. Other processes of higher order are mixed transitions like $\frac{p}{q}\omega_+ \pm \frac{m}{n}\omega_-$ (p, q, m and n are integers). Such mixed processes might also be observable in the discussed LCJJ samples.

It is worth pointing out that the lower resonant mode ω_- might be used for a phase qubit with improved dephasing times compared to a bare Josephson junction. This can be directly seen from the insets in Figure 8.2. Dephasing is mainly caused by low frequency fluctuations in the biasing parameter, which in our case is the bias current. The dephasing rate is given by $\Gamma_\phi \approx \Delta I_n \frac{d\omega}{dI}$, where ΔI_n is the amplitude of the low frequency current noise. We see that a system having a flatter bias current dependence of the resonant mode will be less sensitive to bias current fluctuations. Therefore a large value of β is desirable. However, this value should not be too large. For an increasing value of β, the anharmonicity of the system is reduced, which makes a two-level approximation of the quantum system cumbersome.

8.3 Circuit Design and Simulation

In this experiment, the dynamics of such an LCJJ system is studied in the physically most interesting regime of $\vartheta = C_J/C_s \approx 0.1$ for different $\beta = L_s/L_{J0}$ values. Consequently, only one capacitor but various inductor geometries had to be designed. For the Josephson junctions in the circuits, a diameter of $d \approx 3$ µm and a critical current density of $j_c \approx 80$ A/cm^2 were chosen. This leads to a capacitance of $C_J = 0.4$ pF and a Josephson inductance of $L_{J0} = 0.06$ nH.

8.3.1 Capacitor Design

For the calculation of the shell capacitance values, the simple plate capacitor formula $C_s = \varepsilon_0 \varepsilon_r A / d_i$ was used. Here, ε_0 is the vacuum permittivity, A the capacitor area and d_i the dielectric thickness. Since two insulating layers are used in Josephson junction fabrication, namely Nb_2O_5 and SiO, a system of two capacitors in series has to be considered. Nb_2O_5 was created at an anodization voltage of 20 V, so that a thickness of 46 nm was obtained. The permittivity of this material accounts for $\varepsilon_r = 33$ (see section 4.2.3). SiO with a permittivity of $\varepsilon_r = 5.7$ (see section 4.3.2) was used with a thickness of about 260 nm. Consequently, in order to reach the desired ratio of $C_s/C_J = 10$, an overlay area of $A = 0.02$ mm^2 was chosen for the shell capacitor, so that $C_s = 3.8$ pF.

8.3.2 Inductor Design

The calculation of inductance values is more subtle than for the capacitances. At first, simulations with the free software *FastHenry*[1] were carried out. However, this software only yields values for the geometrical inductance and does not take the kinetic inductance of the superconducting Nb and a potential frequency dependence into account. Consequently, further simulations with *Sonnet*[2], where these contributions can be considered, were performed. We found a significant difference between the obtained inductance values, which indicates that the contribution of the kinetic inductance was important. Furthermore, for some designs a frequency dependence of the inductance was indeed found. In these cases, the design inductance value was taken at a frequency $f = 1/(2\pi\sqrt{(L_s + L_{J0})(C_s + C_J)})$.

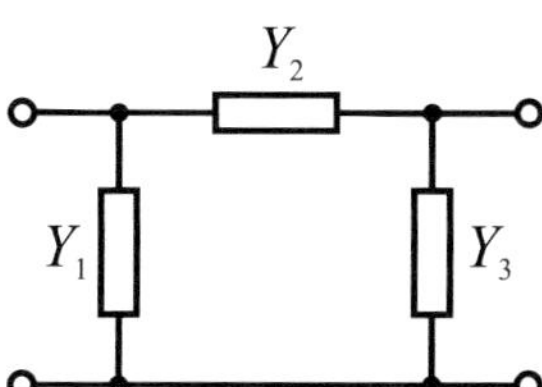

Figure 8.3: General two-port network for simulation of an element of admittance Y_2 between port 1 and 2. This model was used to calculate the inductance values from the *Sonnet* simulation results.

---

[1] Fast Field Solvers, http://www.fastfieldsolvers.com
[2] Sonnet Software Inc., 1020, Seventh North Street, Suite 210, Liverpool, NY 13088, USA

Here, we chose the lower resonant mode as it has a stronger deviation from the bare RCSJ plasma frequency.

The simulations with *Sonnet* were carried out with a typical two-port network. The software is able to give out scattering, impedance or admittance matrices. Since the existence of parasitic elements between the ports and ground has to be assumed, a general network model as shown in Figure 8.3 has to be applied. The admittance matrix of such a network is given by

$$\begin{pmatrix} Y_1 + Y_2 & -Y_2 \\ -Y_2 & Y_2 + Y_3 \end{pmatrix}. \tag{8.6}$$

It can be seen that the admittance of the element of interest is simply given by one matrix element, $Y_2 = -Y_{12} = -Y_{21}$, no matter what the values of the parasitic elements Y_1 and Y_3 are. In general, the admittance Y_2 will be given by

$$Y_2 = \tilde{G} + i\left(\omega C - \frac{1}{\omega L}\right), \tag{8.7}$$

where $\tilde{G}$ is the conductance and C is a possible capacitive element. Assuming that $\omega C \ll 1/(\omega L)$, i.e. that the susceptance is purely inductive, the inductance at the simulation frequency f can be calculated as

$$L = \frac{1}{2\pi f \cdot \mathrm{Im}(Y_{12})}. \tag{8.8}$$

Eight different inductance geometries were designed, ranging from short straight lines to relatively narrow meanders (the narrowest meander can be seen in Figure 8.4a), leading to L_s values from 0.17 nH to 16 nH. This means that we could cover a wide range of $L_\mathrm{s}/L_\mathrm{J0}$ ratios from $\beta \approx 3$ to $\beta \approx 270$. Smaller β values than 3 were unattainable because the value of $L_\mathrm{s} = 0.17$ nH was already reached for the required connection from the capacitor to the Josephson junction. For the narrow meander lines, capacitive coupling between neighboring lines might occur. Hence, the assumption that the imaginary part of the admittance Y_{12} is purely inductive, so that (8.8) is valid (see above), has to be tested experimentally. This will be described in the following section.

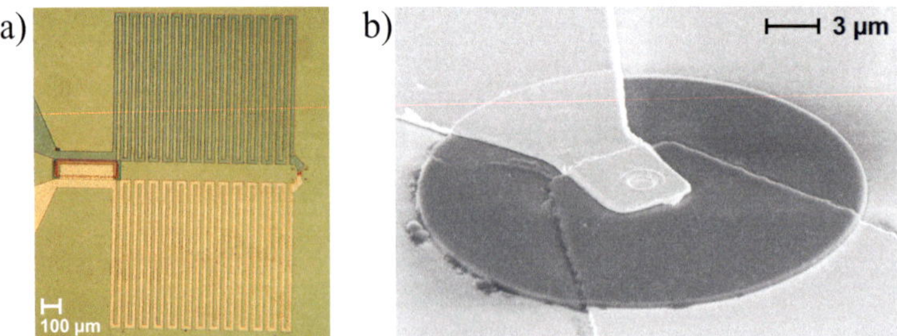

Figure 8.4: a) Photograph of an LCJJ8 type sample. The capacitor is situated on the left while the Josephson junction can be found on the right. The two meander lines connecting these elements act as inductors. The bottom electrode (top of picture) has changed color due to the anodic oxidation. b) SEM image of the Josephson junction itself. The SiO insulation can be seen as the dark circle.

8.4 Sample Fabrication and Pre-Characterization

The samples were fabricated using the combined photolithography / electron-beam lithography process described in section 4.3. However, the Al hard mask technique had not been developed at that time, so that the Josephson junctions were defined using negative e-beam lithography. This was not a problem concerning encroachment (see section 4.2.3), since a relatively large junction diameter of 3 µm was desired anyway. However, the usage of e-beam resist for junction definition lead to a not well-defined junction size. In order to always obtain a capacitance ratio $\vartheta = 0.1$, the junction size was measured for each sample with scanning electron microscopy (SEM) and the SiO thickness was adjusted accordingly. An overview of working samples which exhibited high quality in pre-characterization can be found in Table 8.1.

In order to minimize Nb_2O_5 encroachment (see section 4.2.3), the anodic oxidation was carried out with a voltage of only 20 V. For the circuits, a critical current density of ≈ 80 A/cm^2 was chosen. The inductors and the capacitor plates were simply patterned by photolithography as parts of the bottom electrode M2a and the wiring layer M3. The capacitor dielectric was patterned at same time as the SiO vias by positive electron-beam lithography. A photograph of a typical sample can be seen in Figure 8.4a, while an SEM image of the junction itself is shown in Figure 8.4b.

Josephson Junction Characterization

As a first step, single Josephson junctions with parameters identical to the ones used for the final experiments were fabricated and characterized in order to see if a sufficient quality could be reached. From the measurements at 4.2 Kelvin, the quality factor R_{sg}/R_N was evaluated. For all junctions, we found $R_{sg}/R_N > 35$, which indicates a very high quality. Since the critical currents of the order of 5 µA were already strongly suppressed at 4.2 K, one Josephson junction was investigated in more detail at mK temperatures; its *IV* curve can be seen in Figure 8.5. We found no excess currents and a high $I_c R_N$ product of 1.73 mV,

Table 8.1: Designed parameters of working *LC* shunted Josephson junctions. The difference in junction size is due to the patterning with negative e-beam resist, as the Al hard mask technique had not been developed yet.

Sample	LCJJ1	LCJJ2	LCJJ2$'$	LCJJ3	LCJJ6	LCJJ7	LCJJ8
j_c [A/cm^2]	77	77	77	87	77	77	73
I_c [µA]	5.7	5.0	5.5	6.7	5.0	4.4	3.3
d [µm]	3.06	2.87	3.01	3.12	2.87	2.7	2.39
C_J [pF]	0.39	0.34	0.38	0.41	0.34	0.30	0.24
SiO thickness [nm]	260	260	260	250	260	260	400
C_s [pF]	3.7	3.7	3.7	3.8	3.7	3.7	2.4
ϑ [ ]	0.11	0.09	0.10	0.11	0.09	0.08	0.10
L_{J0} [pH]	58	66	60	49	66	75	100
L_s [nH]	0.17	0.48	0.48	0.8	3.2	6.4	16
β [ ]	2.9	7.3	8.0	16.2	48.4	85.7	159.2

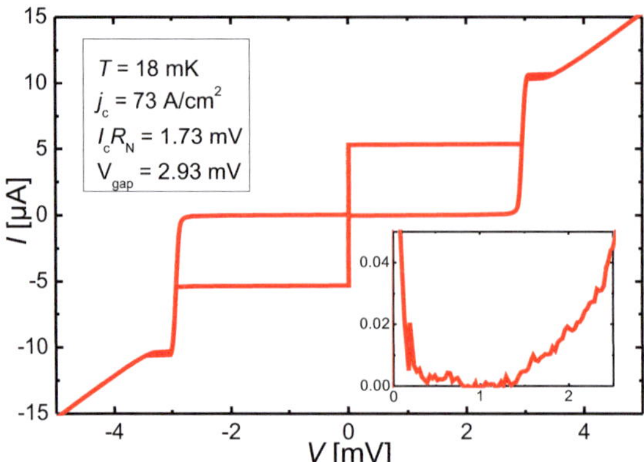

Figure 8.5: *IV* curve of a single junction having a diameter of $d = 3$ µm without the shell circuit. The high I_cR_N and V_{gap} values show that a very high quality was reached. The inset shows a magnification of the subgap branch as measured with a voltage bias, which shows that also very low subgap leakage currents were obtained.

meaning that we observe clear and clean Cooper pair tunneling (see section 3.1.2). Furthermore, measurements with a voltage bias setup showed very low leakage currents also at mK temperatures and voltages close to $V = 0$, as can be seen in the inset of Figure 8.5. The high gap voltage of $V_{gap} = 2.93$ mV indicates that the Nb electrodes are of high quality.

Furthermore, the modulation of the critical current with an external magnetic field was measured and found to be in agreement with the theoretical expectation according to (3.14). This indicates that the bias current is distributed evenly inside the junction. Altogether, we can say that also the junctions used for this experiment (fabricated without the Al hard mask technique) had a very high quality. Before each experiment on an LCJJ circuit, the *IV* curves were measured as well. No significant difference to the *IV* curve of the single junction was found. For sample LCJJ2′, for example, an I_cR_N product of 1.70 mV and a gap voltage of $V_{gap} = 2.92$ mV were observed.

Characterization of the *LC* Shell Circuits

In the following, the characterization of the external shell circuit will be described. In order to see whether the simulation values were correct, we fabricated *LC* shell circuits with the same design as for the final samples, but replaced the Josephson junctions with a superconducting shortcut. Furthermore, a small gap was etched into one of the bias lines leading to the *LC* circuit. In this way, a coupling capacitor C_c was formed, which should sufficiently decouple the *LC* circuit from the low-impedance environment. The experimental test of our designs was especially important for the structures with narrow meander lines, in order to exclude that a spurious capacitive coupling between the lines would shift the resonance or degrade its quality. For this purpose, we designed and fabricated a printed circuit board that allowed measurements up to about 2 GHz, which was enough to test the three smallest resonance frequencies and hence the three largest inductors (details about the design of this PC board can be found in the *Studienarbeit* of Daniel Bruch [Bru08]). For one design, the

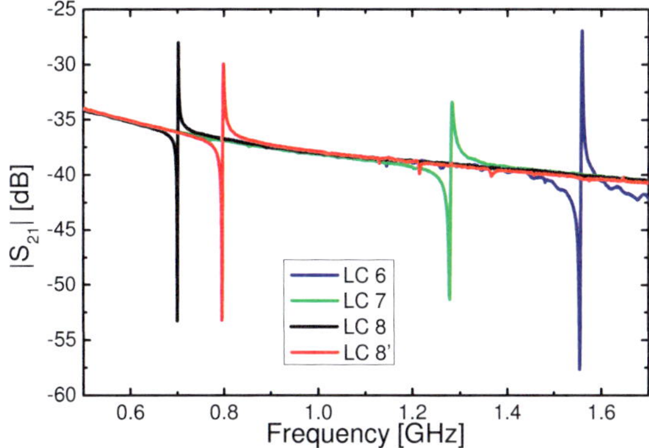

Figure 8.6: Transmission measurement of shell circuits without the Josephson junctions. We found clear resonances at the expected frequencies with high loaded quality factors Q_L (see Table 8.2).

resonance frequency was slightly varied by choosing two different SiO thicknesses, so that two samples LC8 and LC8$'$ were obtained.

These preliminary measurements were carried out in liquid helium at 4.2 Kelvin. The chips were glued onto the printed circuit board and contacted by Al bonding wires. The PC board was contacted via mini-SMP connectors and 50 Ω cables to an Agilent Network Analyzer, which was used to measure the transmission S-parameter $|S_{21}|$. The results of the measurement can be seen in Figure 8.6 and Table 8.2. It is clear that we got an excellent agreement between expected and experimentally determined resonance frequencies. This confirms the precision of our simulation method and excludes capacitive parasitics of any kind. We can conclude that if we have such excellent agreement even for the complex high-inductance structures, the parameters of the low-impedance structures, containing simple straight lines as inductors (i.e. samples LC1 - LC5), should also be well-defined. Furthermore, we obtained high loaded quality factors Q_L, which were probably only limited by the not well-defined C_c values. This means that the shell circuits exhibit low damping, which is crucial for the observation of the underdamped dynamics of the LCJJ systems.

Table 8.2: Measurement results for four investigated shell circuits without junctions. The experimentally determined resonance frequencies are within 3% of the ones calculated from the L_s and C_s design values. Additionally, high loaded quality factors Q_L were obtained.

Sample	$f_{\text{res,design}}$ [GHz]	$f_{\text{res,meas}}$ [GHz]	Q_L [ ]
LC6	1.60	1.56	777
LC7	1.27	1.28	138
LC8	0.72	0.70	506
LC8$'$	0.80	0.80	236

8.5 Setup and Procedure of Measurement

The measurements were carried out at the MC2 (Department of Microtechnology and Nanoscience, Chalmers University of Technology, Gothenburg, Sweden) in an existing dilution refrigerator setup. The setup is shown in Figure 8.7, but shall only be discussed briefly here, since the author of this thesis made no contribution to it. As usual for such quantum experiments, great care was taken to reach a low noise and low disturbance measurement environment. Additional to the filters shown in Figure 8.7, the wiring is fed through *LC* filters at room temperature having capacitance values of $C = 2 - 3$ nF. The copper powder filters placed at base temperature involve a 2 m long wire with a diameter of 50 µm and have a resistance of 15 Ω at room temperature. Generally, the wiring at temperatures below 4.2 K consists of constantan while the wiring at 4.2 K and higher temperatures is made out of copper. The sample could be irradiated by microwaves (MW) with the use of a dipole antenna, which was placed near the sample position.

The measurement procedure is in principle identical to the one described in section 7.3. Additionally, spectroscopy experiments were performed on the LCJJ samples. This was done by permanently measuring switching current histograms and changing the frequency of the applied microwave radiation. If a second switching peak occurred in the histogram, the energy $\hbar\omega_{MW}$ of the microwaves matched the level splitting in the potential well. Consequently, the Josephson phase particle was lifted to a higher energy level where the tunneling rate Γ was sufficiently high to allow the phase particle to escape from the well and enter the running state.

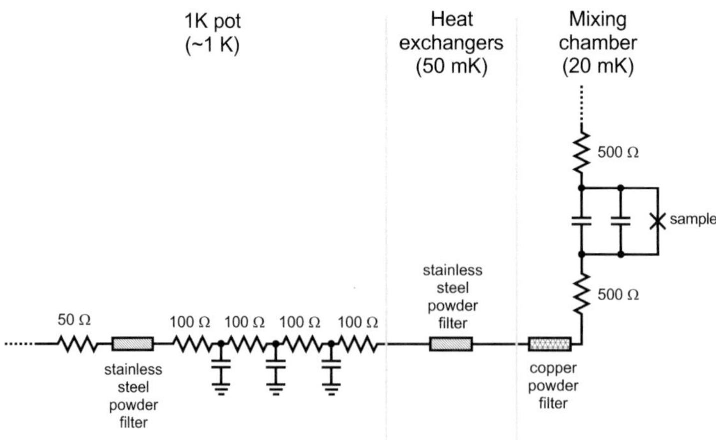

Figure 8.7: Measurement setup for LCJJ measurements at the MC2. All lines are identical and four lines were used per sample. All capacitors have $C = 2$ nF.

8.6 Results and Discussion

Two out of the seven fabricated samples given in Table 8.1 have been measured so far. As has been discussed in section 8.2, the upper mode ω_+ will be very close to the plasma frequency ω_p of the single junction for large β values. Consequently, in order to prove that the observed energy level splittings for the entire LCJJ circuit are indeed different from those of a Josephson junction, the samples with small β values were measured at first. In detail, measurements were started on sample LCJJ2′ and later continued on sample LCJJ1.

8.6.1 Measurements on Sample LCJJ2′

Similar to the measurements discussed in chapter 7, the escape temperature T_{esc} was determined for sample LCJJ2′. In Figure 8.8, T_{esc} is plotted versus the bath temperature T_{bath}. As expected, it is $T_{\text{esc}} \approx T_{\text{bath}}$ in the thermal regime. At lower temperatures, a saturation of T_{esc} can be seen, which can be attributed to the crossover to the quantum regime. However, interpretation of these data is not obvious. For the single junction, the expected crossover temperature can be calculated as $T_{\text{cr,J}} = 115$ mK by evaluating (2.24) at the normalized crossover switching current γ_{cr} determined by (7.1) (for details see chapter 7). But since the dynamics of the single junction should not be observed anymore, the crossover should not occur at this temperature. Here, instead of the escape from a one-dimensional potential well, the escape from a two-dimensional potential well has to be considered. In this case, it is not clear which trajectory the particle will take in the potential landscape, how the corresponding quantum tunneling rate Γ_q has to be calculated and where the crossover from the thermal to the quantum regime will take place. To make predictions for such a system, investigations by specialized theory groups will have to be carried out in the future. Within this thesis, a physical explanation for the experimentally observed behavior cannot be given. In a naive approach, it was tried to simply replace ω_p in (2.24) by ω_- and ω_+ and calculate two new crossover temperatures. For this, the γ_{cr} value determined for the

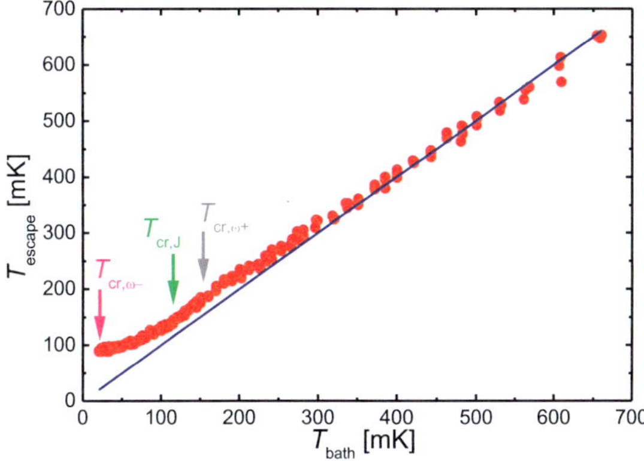

Figure 8.8: Escape temperature of sample LCJJ2′. While the behavior in the thermal regime meets the expectations, the theoretical background of the crossover to the quantum regime is still unclear.

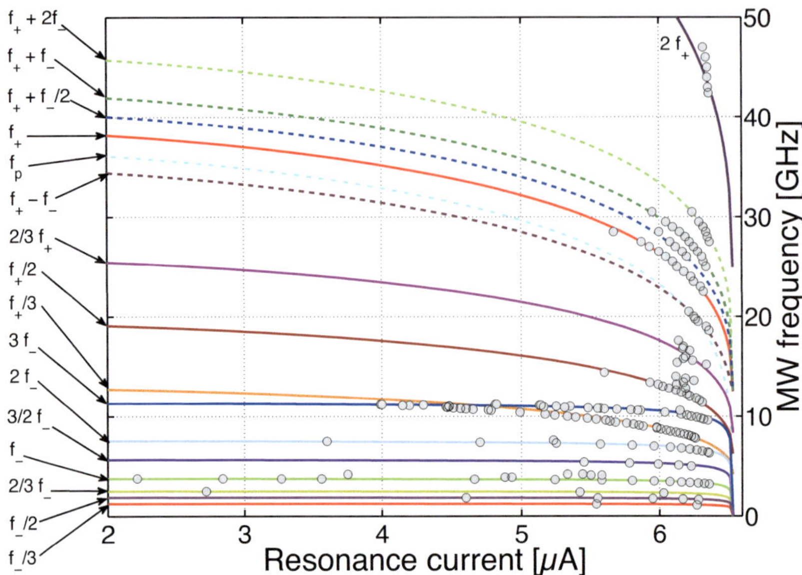

Figure 8.9: Spectroscopy of sample LCJJ2$'$ without applied magnetic field. An excellent agreement between experiment and design values as well as theory is observed.

single junction was taken, which is certainly not correct. The two crossover temperatures $T_{\mathrm{cr},\omega_-} = 20$ mK and $T_{\mathrm{cr},\omega_+} = 154$ mK are marked in Figure 8.8. It can be seen that these naively calculated values do not describe the behavior of the LCJJ system and that further theoretical studies are indeed required.

Spectroscopy measurements were also carried out on sample LCJJ2$'$ in order to reveal whether the two orthogonal plasma modes ω_- and ω_+ would be observed instead of the plasma frequency of the single junction. The measurement results are shown in Figure 8.9. In the same graph, theoretically calculated curves according to (8.4) and (8.5) are shown. It can be seen that besides the normal modes ω_- and ω_+, higher order processes as well as mixed transitions (as discussed in section 8.2) were found. Altogether, all measurement data are in excellent agreement with the theoretically calculated curves. Furthermore, no data points at the plasma frequency of the single junction were found. This means that the LCJJ system is very well described by the theory given in section 8.2 and acts indeed as one single quantum system. This is remarkable since the circuit has a size of 200×650 µm^2. Although no superpositions of quantum states were investigated here, it can be said that we are probably dealing with a relatively large Schrödinger's cat.

All theoretical curves in Figure 8.9 were calculated using only one fixed set of parameters. For the junction, these were a critical current of $I_{\mathrm{c}} = 6.53$ µA and a Josephson capacitance of $C_{\mathrm{J}} = 0.38$ pF. The latter corresponds to a specific capacitance of $c = 53$ fF/µm^2, which is in excellent agreement with the value of of $c = 55$ fF/µm^2 used in chapter 7. Since the critical current density in this experiment is about 10 times lower than in the MQT measurements discussed in chapter 7, the tunneling barrier is thicker here, so that a slightly lower specific capacitance can indeed be expected. For the shell circuit, the parameters

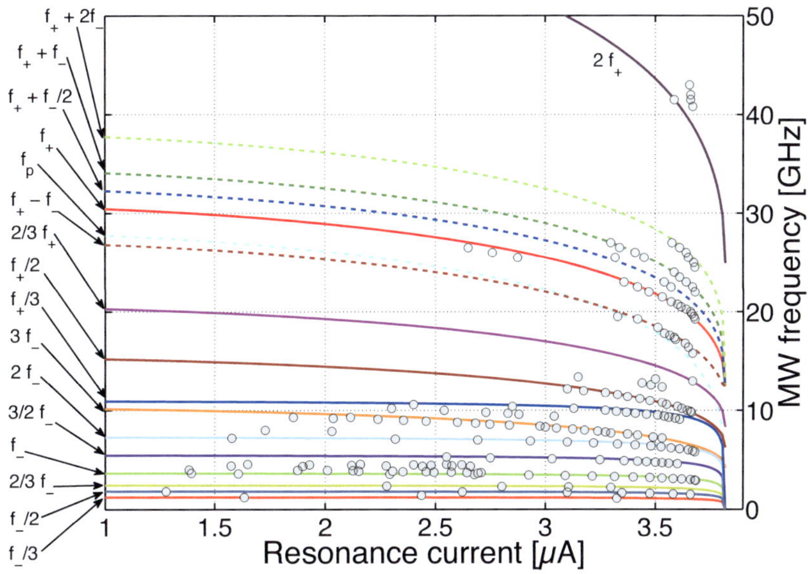

Figure 8.10: Spectroscopy of sample LCJJ2$'$ with applied magnetic field, so that the critical current is suppressed to $I_c = 3.82$ µA. The agreement between theory and experiment is excellent.

accounted for $L_s = 0.43$ nH and $C_s = 3.7$ pF. That means that effectively, $\vartheta = 0.10$ and $\beta = 8.5$. A comparison between the values used for calculation and the design values given in Table 8.1 shows an excellent agreement. Only the critical current was slightly higher and the shell inductance slightly lower than planned. This shows that the parameters of our Josephson junction circuits are very well under control and the quantum properties of the circuits can be designed at will.

On the same sample, the critical current was suppressed to $I_c = 3.82$ µA by applying a magnetic field, and the same spectroscopy measurement was repeated. The measurement results are shown in Figure 8.10. All theoretical curves in this graph were calculated with the same parameters as in the absence of magnetic field. It can be seen that again, an excellent agreement between measurement data and theoretical expectation was reached. This confirms the conclusions discussed above.

8.6.2 Measurements on Sample LCJJ1

Similar spectroscopy measurements were also carried out on sample LCJJ1 in order to confirm the conclusions obtained so far. The measurement data can be seen in Figure 8.11 and show excellent agreement with the theoretical expectation. Again, all theoretical curves were obtained with one fixed set of parameters. Here, these were: $I_c = 6.73$ µA, $C_J = 0.39$ pF (again corresponding to $c = 53$ fF/µm^2), $L_s = 0.19$ nH and $C_s = 3.7$ pF. That means that effectively, $\vartheta = 0.11$ and $\beta = 4.0$. Just as for sample LCJJ2$'$, all values are in excellent agreement with the design values from Table 8.1. Consequently, we can say that

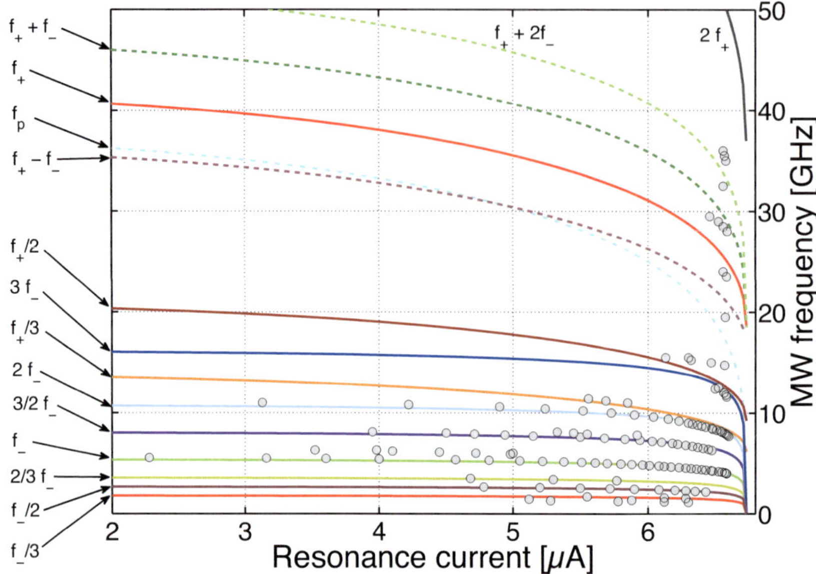

Figure 8.11: Spectroscopy of sample LCJJ1 at zero magnetic field. As for sample LCJJ2′, an excellent agreement between experiment and theory was obtained.

all conclusions given above are confirmed on the second sample. Also for circuit LCJJ1, measurements with an applied magnetic field were performed. In this case, the comparison with the theoretical curves calculated with the same parameters as for zero field also give a reasonable agreement.

8.7 Conclusions

In this chapter, it was shown that a Josephson junction shunted by an inductance and a capacitance acts indeed as one single quantum system. This is remarkable since the circuits have sizes up to $200 \times 650\mu\text{m}^2$, which is larger than for most superconducting qubits. We found an excellent agreement between the theoretically expected behavior calculated using the design parameters and the measurement data. The findings show that such an LCJJ system has indeed two new energy scales instead of the energy scale of a single Josephson junction. Both plasma modes were observed in the investigated samples, which has never been reported before. Our results are important for the understanding of the dynamics in high-temperature Josephson junctions, where such shunting elements cannot always be avoided in fabrication. Furthermore, the findings are of great interest for superconducting quantum circuits such as quantum bits. When such systems are shunted capacitively with a shell capacitance C_s, it is mostly assumed that $L_s = 0$ (which is definitively not the case since all electrical connections exhibit an inductance) and $\omega_p = \omega_- = 1/\sqrt{L_J C_s}$. We have shown that L_s needs to be considered resulting in $\omega_- = 1/\sqrt{(L_J + L_s)C_s}$ and that there is a second energy scale in the system, namely $\omega_+ = \sqrt{(1/L_J + 1/L_s) \cdot 1/C_J}$. This is an important result for quantum device design and operation.

9 Coherent Oscillations in Nb/Al-AlO$_x$/Nb Phase Qubits

In the previous chapters, it has been shown that the Josephson junctions fabricated with the newly developed technological process indeed behave as macroscopic quantum systems. Consequently, this technology was used to build a quantum device, i.e. a superconducting qubit. Since the quantum dynamics in such a system should evolve coherently for as long as possible, the technological requirements are even harder than for other quantum experiments such as MQT. Consequently, this chapter aims at fabricating phase qubits with long coherence times. In order to contribute to a better understanding of the decoherence processes in superconducting phase qubits, the measured coherence times shall be related to the employed fabrication process.

This chapter begins with a short introduction discussing the special technological requirements for phase qubits. Second, the the sample design including all circuit elements is presented. Then, the dynamics of a quantum mechanical two-level system are discussed, which includes the typical measurements for determination of the coherence times. Finally, the measurement results are presented and evaluated concerning the influence of the technological process on the qubit coherence times. Some parts of this chapter have been published in [KMI$^+$11].

9.1 Introduction

Quantum computers can solve certain problems exponentially faster than any classical computer, such as the simulation of quantum systems [18], the prime factorization of large numbers [19] or searching an unsorted database [20]. As the constituting element of such a quantum computer, a quantum bit, any quantum mechanical two-level system can be taken as long as the following criteria presented by DiVincenzo [17] are fulfilled:

- A scalable physical system with well characterized qubits
- The ability to initialize the state of the qubits
- Long relevant coherence times, much longer than the gate operation time
- A universal set of quantum gates
- A qubit-specific measurement capability

While natural quantum systems such as atoms, ions, nuclear spins etc. exhibit very long coherence times, it is difficult to control and scale such systems to the size of a quantum computer. Furthermore, their characteristic parameters are fixed by nature. For a superconducting qubit, the opposite is the case: The parameters can be designed at will and the qubits are fabricated with chip technology, so that they are easily scalable. On the other hand, the coherence times are severely limited and do currently not exceed ≈ 1 µs for any of the various types of superconducting qubits.

The reasons for these limited coherence times are not *a priori* clear. For phase qubits (see section 2.4), which are investigated in this chapter, it was found recently that the presence of spurious two-level systems (TLS) in the vicinity of these devices limits their coherence times [23, 84]. These TLS were found in dielectric thin films [23, 94] as well as on metal surfaces [88, 95]. It has been found that the coherence times in qubits involving Al/AlO$_x$/Al junctions are significantly longer than in qubits fabricated with Nb/AlO$_x$/Nb technology [21, 22]. Since the decoherence rate due to subgap leakage should not be significant in either case [64], the difference in coherence times is related to a lower TLS density in Al based phase qubits. However, it still remains unclear where exactly the TLS in Nb based qubits come from and how they can be eliminated.

9.2 Sample Design

9.2.1 The Phase Qubit

The basic principle of phase qubit operation is sketched in Figure 2.6. The ground state $|0\rangle$ and the excited state $|1\rangle$ of the qubit are given by the two lowest energy states in the potential well. Consequently, in order to be able to design the qubit and predict the microwave frequency needed for quantum manipulation, the energy eigenstates E_n in the potential well need to be calculated. How this was done within this thesis is described in detail in Appendix D.

For the qubit design, two requirements have to be fulfilled for the energy separation Δ_{01} between the ground and the excited state. First, it should be $\Delta_{01} \gg k_B T$, so that the qubit will not be excited thermally. For a qubit operation temperature of $T = 100$ mK, the thermal energy would account for ≈ 2 GHz (since in qubit research, manipulation is carried out by microwaves (MW), all energies are divided by Planck's constant h and given in GHz).

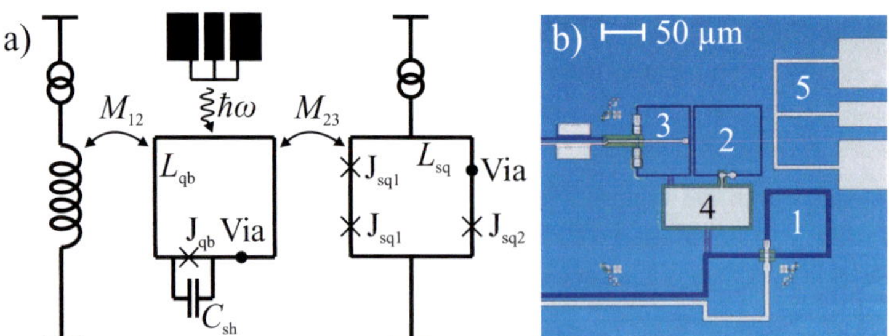

Figure 9.1: a) Circuit schematics of the investigated phase qubits. Qubit Ch2 did not have a shunt capacitance C_{sh}. All vias necessary for the connection of the wiring to the lower electrode were not realized as large junctions, but the tunneling barrier was etched away with IBE, so that real vias were formed. The designed circuit parameters are given in Table 9.1. b) Photograph of the sample Ch1 with 1 = Bias coil, 2 = Phase qubit, 3 = DC-SQUID, 4 = C_{sh}, 5 = MW line.

There is also an upper limit for Δ_{01}, as the MW lines and attenuators in a typical dilution refrigerator are not well-defined for higher frequencies. Experimental difficulties can be expected above 15 GHz while above 20 GHz, the microwave power reaching the qubit might not be sufficient anymore to carry out the desired manipulations. Consequently, typical values for the qubit level splitting lie in the range of $\Delta_{01} = 6 - 8$ GHz. Additionally, the value of Δ_{01} should differ significantly from the next higher level splitting Δ_{12} (in other words, a sufficient anharmonicity should be ensured), so that the qubit can be driven selectively between the states $|0\rangle$ and $|1\rangle$ without exciting it to higher energy levels.

Furthermore, care should be taken to obtain a reasonable value for the qubit screening parameter $\beta_{L,\mathrm{qb}}$ introduced in (2.35):

- For $\beta_{L,\mathrm{qb}} < 1$, the qubit potential (2.36) has only one minimum, so that qubit operation in a metastable potential well as described in section 2.4 is not possible. Hence, the states in the potential well cannot be read out, so that such a system cannot be used as a qubit.

- For $1 < \beta_{L,\mathrm{qb}} < 4.6$, the potential (2.36) has exactly two local minima, so that qubit operation can be carried out as depicted in Figure 2.6: Qubit manipulation is performed in the upper, metastable well. Then, for read out, the potential is tilted so that the excited state tunnels to the lower well, which is equivalent to one flux quantum entering into the qubit loop. Through this difference in magnetic flux, the qubit can be read out by a DC-SQUID.

- For $\beta_{L,\mathrm{qb}} > 4.6$, the qubit potential has more than two local minima. While in principle, a qubit operation as just described can still be carried out, the ambiguity makes qubit initialization and data analysis more difficult.

For qubit design, we chose a trilayer with a rather low critical current density $j_\mathrm{c} = 65$ A/cm^2. Furthermore, the specific junction capacitance $c = 53$ fF/μm^2 has a fixed value (the same value for c was taken for the experiments in chapter 8, where a similar critical current density was used). Consequently, the only adjustable parameters are the qubit self-inductance L_qb (which is defined by the loop geometry and was simulated with the free software *FastHenry*[1]), the junction diameter $d_\mathrm{J,qb}$ and a potential shunt capacitance C_sh (see Figure 9.1a). In the qubit design process, these free parameters must be varied and the corresponding Δ_{01} and β_L values have to be calculated, in order to see if the discussed requirements are fulfilled.

The designed phase qubit parameters can be found in the upper part of Table 9.1. We decided to fabricate one qubit with shunt capacitance and one without. Compared to the qubit without C_sh, the electric field is redistributed from the qubit junction to the capacitance in a qubit with shunted design. Consequently, a difference in coherence times between such structures might reveal different TLS densities in the junction and the capacitor. However, such a comparison is probably difficult to carry out since the level splitting Δ_{01} differs strongly if all other parameters are kept constant (see Table 9.1). For the designed samples, the value of Δ_{01} is in a reasonable range for the qubit with shunt capacitor, while qubit manipulation might already be difficult for the sample without additional capacitance. The screening parameters $\beta_{L,\mathrm{qb}} = 3.8$ should result in non-ambiguous qubit operation.

---

[1]Fast Field Solvers, http://www.fastfieldsolvers.com

Table 9.1: Design parameters of the investigated circuits containing phase qubits. The value of j_c was measured on test junctions on the same trilayer while all inductances and the level splitting Δ_{01} were calculated numerically.

	Sample Ch1	Sample Ch2
L_{qb} [pH]	257	257
j_c [A/cm^2]	65	65
$d_{J,qb}$ [µm]	3.1	3.1
$I_{c,qb}$ [µA]	4.9	4.9
$\beta_{L,qb}$ [ ]	3.8	3.8
C [pF]	0.4	0.4
C_{sh} [pF]	0.82	-
Δ_{01} [GHz]	8.8	17.8
L_{sq} [pH]	227	227
$d_{J,sq1}$ [µm]	2.83	2.83
$d_{J,sq2}$ [µm]	2.0	2.0
$I_{c,sq}$ [µA]	4.1	4.1
β_L [ ]	2.8	2.8
Modulation depth [%]	23	23
M_{12} [pH]	1.36	1.36
M_{23} [pH]	22.2	22.2
M_{13} [pH]	0.12	0.12
$\Delta\Phi_{sq}$ [mΦ_0]	53	53

It should be mentioned that the calculation of the energy levels and hence the obtained Δ_{01} and Δ_{12} values are not exact for the qubit junction shunted by an external capacitor. As has been shown in chapter 8, the inductive lines between the junction and the shunt capacitor lead to a two-dimensional potential with two energy scales. While the equations for the two normal modes ω_- and ω_+ are given in (8.4) and (8.5), a precise calculation of the energy levels in the shallow two-dimensional potential well including determination of the anharmonicity would require solving the corresponding Schrödinger equation (D.1). This is far more complicated than for a one-dimensional potential and has not been carried out within this thesis. However, the qubit level splitting should only fulfill the requirements described above and does not have to be known precisely for qubit design. Consequently, the qubit capacitance C was simply replaced by $C + C_{sh}$ in the case of capacitive shunting and the calculation of the energy levels was carried out with a one-dimensional potential.

9.2.2 The DC-SQUID

In order to obtain a clear read out signal, the DC-SQUID should have a sufficient modulation depth. The latter depends on the value of the screening parameter β_L in (2.33), as discussed in section 2.3. Usually, a small β_L value is reached by reducing the inductance of the SQUID loop L_{sq}. Here, this is not possible since the coupling to the qubit has to have a certain strength. Instead, the critical current of the SQUID junctions can be made small. First, a low critical current density $j_c = 65$ A/cm^2 was chosen for the circuit anyway. Second, the qubit contains rather small junctions, so that the same can be done for the SQUID. Hence, the latter will indeed have a rather small critical current and the modulation depth will be reasonable. The designed SQUID parameters can be found in the middle part of Table 9.1. The modulation depth of 23 % should result in a good SQUID read out resolution.

It can also be seen in Table 9.1 and Figure 9.1 that one arm of the SQUID contains two junctions with a diameter $d_{J,sq1}$, while the other arm contains only one junction with a diameter $d_{J,sq2}$, with $2d_{J,sq2}^2 = d_{J,sq1}^2$. This was done to obtain an asymmetric SQUID modulation curve [119], which has two advantages: First, the side of the curve with a steeper $I_c(\Phi)$ slope can be used in order to get a better read out resolution. Second, the maximum of the SQUID's critical current is shifted from zero magnetic flux, so that no extra flux bias is required to move the SQUID to a working point with steeper slope.

9.2.3 Manipulation and Read Out

In order to ensure reliable qubit manipulation and read out, several bias elements and coupling strengths have to be chosen carefully. The least critical part is the design of the microwave (MW) lines and their coupling to the qubit. Due to all the attenuators at different temperature stages in a dilution refrigerator, the MW power arriving on the chip is not precisely known. Furthermore, the required MW frequency is not known in advance, as it differs from qubit to qubit and can change strongly due to technological deviations. However, knowledge of the frequency would have to be known for an optimized MW line design. Consequently, a thorough design of the coupling strength to the qubit is not necessary and the MW power can just be increased until qubit manipulation is observed.

The coupling of all other elements in circuit is of inductive nature and should be designed carefully. This was done by simulating the geometrical inductances of qubit, SQUID and bias coil with *FastHenry*. The bias coil was realized in a gradiometric design, in order to obtain strong coupling to the qubit M_{12} but only weak coupling to the SQUID M_{13}. As can be seen in the lower part for Table 9.1, this was indeed achieved with $M_{12}/M_{13} = 11$.

In order to obtain a clear flux signal in the SQUID, its coupling to the qubit M_{23} should be rather large. On the other hand, too strong coupling might allow noise in the SQUID to couple into the qubit and decrease its coherence times. As this was the first qubit design at the IMS, a rather strong coupling was chosen to make sure that a successful read out would be possible. The design values $M_{23} = 22$ pH and $I_{c,qb} = 4.9$ µA lead to an expected qubit signal of $\Phi_{sq,qb} = M_{23} \cdot I_{c,qb} = 53$ mΦ_0 in the SQUID. This should result in a clear read out signal, but might be disadvantageous for the observed coherence times. But since we expect the decoherence in phase qubits to be mainly due to TLS in their vicinity, the coupling strength to the SQUID should not play a crucial role in this context.

Analogous to the estimation of the decoherence rate in a single Josephson junction due to quasi-particle leakage via the ringdown time $T_1 = RC$ (see section 3.1.3), such a ringdown time can be estimated for the entire circuit. Since the phase qubit is decoupled inductively from the bias lines with impedance $Z_0 \approx 100\,\Omega$, an impedance transformation according to

$$Z_{\text{eff}} = \frac{L_{qb}}{M_{12}} Z_0 \qquad (9.1)$$

is obtained. Multiplying this effective impedance with the qubit capacitance C yields the ringdown times $T_1 \approx 2.2$ µs for qubit Ch1 and $T_1 \approx 0.7$ µs for qubit Ch2. This is two orders of magnitude longer than the measured T_1 times in any Nb based phase qubit (see section 9.5.3) and even longer than in the best Al based phase qubits. Consequently, it can be assumed that this decoherence mechanism will not play a role.

9.3 Dynamics of a Quantum Mechanical Two-Level System

In order to understand the qubit measurement procedures and the determination of the qubit coherence times, the coherent dynamics in a quantum mechanical two-level system have to be explained. In the following, only the two lowest energy levels $|0\rangle$ and $|1\rangle$ in the phase qubit potential well are considered, so that an effective two-level system is obtained.

In contrast to a classical system, the quantum system can be in superpositions of the basis states like

$$|\psi\rangle = c_0 |0\rangle + c_1 |1\rangle . \tag{9.2}$$

Here, c_i are complex numbers. When a measurement is performed, the general (pure) qubit state $|\psi\rangle$ is projected onto one of the basis states, so that $|0\rangle$ is measured with a probability $p_0 = |c_0|^2$ and $|1\rangle$ is measured with a probability $p_1 = |c_1|^2$. These probabilities can be reconstructed by carrying out the measurement many times. Since one of the basis states is always measured, $|c_0|^2 + |c_1|^2 = 1$ has to be fulfilled. The probabilities p_0 and p_1 are often called the occupations of the states $|0\rangle$ and $|1\rangle$, respectively. It is very intuitive and also helpful to describe the qubit state in spherical coordinates (see Figure 9.2). Since the sum of the probabilities to measure the states $|0\rangle$ and $|1\rangle$ has to be one, the radius will account for one and only the angular coordinates θ and ϖ have to be used. If the prefactor of $|0\rangle$ is chosen to be real, an arbitrary pure state of the qubit can be described as

$$|\psi\rangle = \cos(\theta/2) |0\rangle + e^{i\varpi} \sin(\theta/2) |1\rangle . \tag{9.3}$$

This picture is equivalent to that of a spin-$\frac{1}{2}$ particle, so that the phase qubit can also be called a pseudospin. The azimuthal angle θ turns the qubit between the two basis states while the polar angle ϖ describes rotation around the quantization z-axis. Like for a spin, the qubit vector precesses around the quantization axis with the Larmor frequency

$$f_{01} = \Delta_{01}/h , \tag{9.4}$$

which is simply determined by the energy difference between the ground and the excited state. For a controlled manipulation of the pseudospin, the acting field should always be perpendicular to it. Due to its precession, a description of the qubit vector and its manipulation is rather complicated. It can be simplified by looking at the Bloch vector in the "rotating frame", i.e. in a coordinate system rotating with the Larmor frequency f_{01}. In the rotating frame, no precession of the pseudospin is present, so that it can be described as a constant vector in the Bloch sphere. Using a microwave which is a cosine of frequency $f_{MW} = f_{01}$ for qubit manipulation results in a constant field, which is always perpendicular

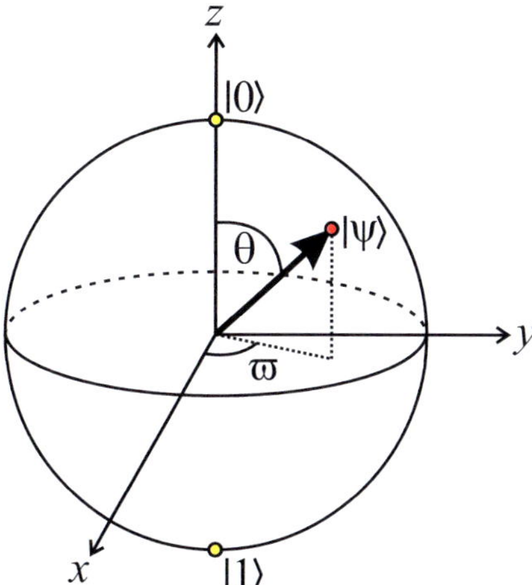

Figure 9.2: Bloch sphere description of the qubit state. The yellow points indicate the basis states |0⟩ and |1⟩ and the red point indicates an arbitrary pure state |ψ⟩ described by (9.3).

to the pseudospin, in the rotating frame. Hence, the qubit vector is rotated around θ by the microwave, with the value of θ proportional to the length and the power of the MW pulse.

9.3.1 Characteristic Measurements and Forms of Decoherence

Irradiation of the qubit by microwaves allows a controlled movement of the pseudospin on the Bloch sphere. If no field is applied, the pseudospin should remain stable in its position in the rotating frame. However, due to different decoherence mechanisms, the position of the pseudospin might change or be smeared out.

Rabi Oscillations

For a Rabi experiment, as for all qubit experiments, the qubit has to be initialized in its ground state |0⟩ at first. Then, the qubit is irradiated with a microwave pulse of frequency f_{01}, constant power P_{MW} and increasing pulse length t_{MW}. Right after the MW pulse is over, the qubit is read out as described in section 9.4. The resulting angle θ is directly proportional to the length of the MW pulse like $\theta = \omega_{\mathrm{Rabi}} \cdot t_{\mathrm{MW}}$. Since the occupation of the ground state |0⟩ is given by $p_0 = |c_0|^2 = \cos^2(\theta/2)$, it will oscillate like

$$p_0 = \cos^2\left(\frac{\omega_{\mathrm{Rabi}} \cdot t_{\mathrm{MW}}}{2}\right) = \frac{1}{2} + \frac{1}{2}\cos(\omega_{\mathrm{Rabi}} \cdot t_{\mathrm{MW}}), \qquad (9.5)$$

with the Rabi-frequency given by [120]

$$\omega_{\mathrm{Rabi}} = \frac{I_{\mathrm{MW}}}{2}(2\Delta_{01}C)^{-\frac{1}{2}}. \qquad (9.6)$$

Here, I_{MW} is the microwave amplitude and C the qubit capacitance. Due to different decoherence mechanisms, the Rabi oscillations will decay exponentially with a characteristic time T_d, which is ideally given by the mean of the relaxation time T_1 and the dephasing time T_2. These will be defined in the following. For stronger driving (i.e. for higher MW power), higher energy levels than $|1\rangle$ in the qubit potential well are also excited, so that (9.6) is not valid anymore. Including transitions to the second excited qubit state $|2\rangle$ leads to an expected oscillation frequency [120, 121]

$$\Omega_\mathrm{R} \propto I_\mathrm{MW} \left(1 - \varkappa I_\mathrm{MW}^2\right) . \tag{9.7}$$

The measurement of Rabi oscillations can be used to calibrate the MW pulse length for constant microwave power. For a certain t_MW value, the value of p_0 will have gone from 0 to 1 for the first time. This means that the pseudospin has been rotated by $\theta = \pi$ (in other words: the qubit has been brought from its ground to its excited state), so that such a pulse is called a π-pulse. If a pulse of half this length is applied (i.e. a $\pi/2$-pulse), the qubit is brought from the ground state to the equatorial plane of the Bloch sphere. This knowledge is very helpful to carry out qubit operations.

Qubit Relaxation

One form of decoherence is relaxation from the excited to the ground state. Since $|1\rangle$ is of higher energy than $|0\rangle$, the qubit will lose energy and relax from $|1\rangle$ to $|0\rangle$ if mechanisms of dissipation are available. This can be measured by initializing the qubit, applying a π-pulse and waiting a time t_delay until the qubit is read out. While the qubit will be in the excited state right after the π-pulse, it will decay to the ground state exponentially with a characteristic time T_1 like

$$p_1 = \exp\left(-\frac{t_\mathrm{delay}}{T_1}\right) . \tag{9.8}$$

T_1 is commonly known as the qubit relaxation time.

Ramsey Experiments and Dephasing

While relaxation describes unwanted evolution of the angle θ, dephasing describes uncontrolled evolution of the angle ϖ. Since states for constant θ but different ϖ do not differ in energy, the value of ϖ does not relax down to a certain point on the Bloch sphere. Instead, the value of ϖ will be more and more smeared out, so that the pseudospin is not evolving coherently anymore and the outcome of an experiment will be unpredictable. The characteristic time for this is called the dephasing time T_2.

T_2 can be measured by initializing the qubit, applying a $\pi/2$-pulse, waiting for a time t_delay, applying another $\pi/2$-pulse and reading out the qubit state right afterwards. If no dephasing was present, the two $\pi/2$-pulses would always bring the qubit in its excited state, so that $p_1 = 1$ would be obtained. After dephasing has occured, however, the position of the pseudospin will not be well-defined anymore, so that the second $\pi/2$-pulse will turn the Bloch vector by $\pi/2$ around an arbitrary rotation axis. Consequently, any qubit state might be measured, so that in average, $p_1 = \frac{1}{2}$ will be obtained. In summary, the occupation of the excited state will evolve with the delay time like

$$p_1 = \frac{1}{2} + \frac{1}{2}\exp\left(-\frac{t_\mathrm{delay}}{T_2}\right) . \tag{9.9}$$

Such an exponential decay from 1 to $\frac{1}{2}$ has limited visibility and is not easily identified as coming from a real physical process. Hence, a more elegant way is often used to determine T_2: the measurement of Ramsey fringes. For this, the microwave frequency f_{MW} is slightly detuned from the Larmor frequency f_{01}. This means that in the rotating frame, the orientation of the qubit manipulation field rotates with a frequency $|f_{01} - f_{MW}|$. For the same experiment as described above, the second $\pi/2$-pulse will not necessarily rotate the pseudospin to the $|1\rangle$-state, but to a certain, well defined point on the Bloch sphere, depending on how much time t_{delay} has passed. Consequently, the occupation of the excited state will exhibit a decaying oscillation like

$$p_1 = \exp\left(-\frac{t_{delay}}{T_2}\right) \cos\left[2\pi(f_{01} - f_{MW})t_{delay}\right] . \tag{9.10}$$

Such an experiment can be carried out for many different values of detuning and the average of all determined T_2 values can be taken as the dephasing time.

9.4 Setup and Procedure of Measurement

For the phase qubit measurements, the existing dilution refrigerator setup of the group of A. V. Ustinov at the Physikalisches Institut was used. Since no contribution to the measurement setup was made by the author, it shall not be discussed here. Details about it can be found in the PhD thesis of Jürgen Lisenfeld [120].

In the precedent section, some qubit measurement procedures were already explained schematically in the Bloch sphere picture. Now, a general, experimental qubit measurement

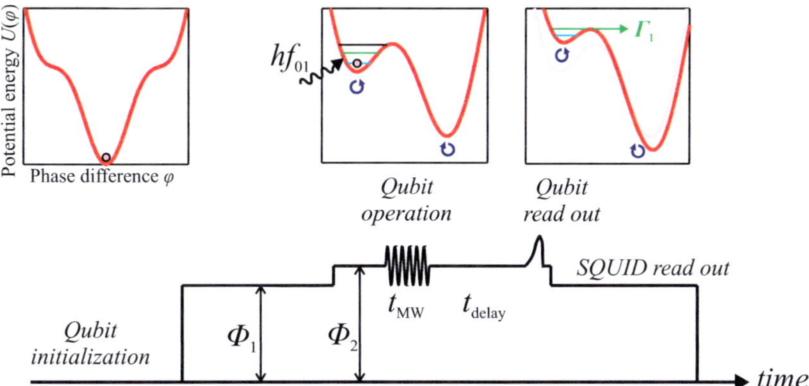

Figure 9.3: Procedure of phase qubit measurements. First, the qubit is initialized in the ground state of the potential well. Then, the potential is tilted, so that the qubit operation can be carried out by MW pulses. To map the qubit states to the different potential wells (corresponding to opposing circulating currents in the qubit loop), a short DC pulse is applied. Right after the read out pulse, the qubit state is frozen at a lower potential tilting angle to allow for a reliable SQUID read out. The SQUID detects the flux signal created by the circulating current in the phase qubit.

procedure will be discussed in more detail. An overview can be seen in Figure 9.3. At first, the qubit has to be initialized. This is done by tilting the potential to the point where only one minimum exists. Next, the qubit is prepared for operation by adiabatically applying a DC magnetic flux Φ_1 in the bias coil, so that the qubit will be in a shallow, metastable potential well. For qubit operation, the working point is reached by a magnetic flux Φ_2, which is slightly larger than Φ_1. Here, two conditions must be fulfilled: First, the tunneling rates from the ground state and the excited state have to be negligible, i.e. $\Gamma_0 = 0$ and $\Gamma_1 = 0$, and second, the anharmonicity in the well must be large enough to make sure that only transitions between $|0\rangle$ and $|1\rangle$ are induced by the resonant microwave pulse. At the working point, qubit operations are carried out as described in section 9.3.1.

For read out, the potential is tilted very shortly to a point where tunneling from $|1\rangle$ is occurring, so that $\Gamma_1 \gg 0$, while still $\Gamma_0 \approx 0$. This is done by application of a short (1 ns) DC flux pulse. After this read out pulse is over, the qubit will by in the lower well if its state was $|1\rangle$ before, but remain in the shallow upper well if it had been in $|0\rangle$. In other words, the qubit states have been mapped to the different potential wells. Since these wells correspond to circulating currents in opposite direction, tunneling to the deeper well involves tunneling of one magnetic flux quantum into the qubit loop, which is detected by the inductively coupled DC-SQUID. During SQUID read out, the qubit potential is brought back to the tilt angle given by flux Φ_1 to make sure that no further tunneling in the qubit can occur.

The flux in the SQUID loop is given by the externally applied flux in addition to the flux signal coming from the qubit. Depending on the qubit state before read out, this qubit signal differs by $\Phi_{\mathrm{sq,qb}} = M_{23} \cdot I_{\mathrm{c,qb}}$. Consequently, two different critical currents $I_{\mathrm{c}}(\Phi_{\mathrm{qb}=|0\rangle})$ and $I_{\mathrm{c}}(\Phi_{\mathrm{qb}=|1\rangle})$ are obtained for the SQUID. Since every qubit measurement has to be carried out a large number of times to reconstruct the occupations p_0 and p_1 anyway, two separate full switching current histograms are obtained for the SQUID (see Figure 9.4). By taking the weight of these switching peaks, the escape probability of the qubit from the shallow well P_{esc} can be calculated. Ideally, it would simply be $P_{\mathrm{esc}} = p_1$. In the actual experiments,

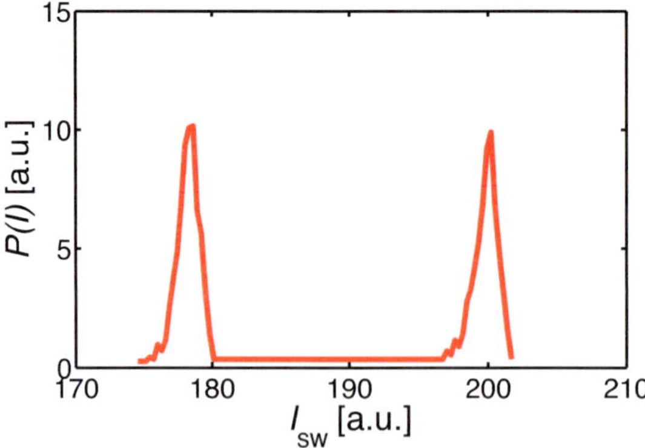

Figure 9.4: The two separate SQUID switching peaks for a qubit level occupation of $p_0 = p_1 = \frac{1}{2}$ in sample Ch1. The left histogram corresponds to the qubit state $|0\rangle$ while the right histogram belongs to $|1\rangle$. The distance of the peaks can be used to determine the mutual inductive coupling between qubit and SQUID, as described in the text.

however, the flux Φ_2 was calibrated to $P_{\mathrm{esc}} \approx 0.1$ even for the qubit in the ground state, in order to obtain a high anharmonicity. Consequently, $p_1 \neq P_{\mathrm{esc}}$ and a maximum visibility in P_{esc} of 0.8 can be reached.

9.5 Results and Discussion

As can be seen in Table 9.1, two different qubits were characterized: one with shunting capacitor (Ch1) and one without (Ch2). The results for these samples will be discussed separately.

9.5.1 Characterization of Phase Qubit Ch1

First, by measuring the distance of the two separate SQUID switching peaks (see Figure 9.4) and taking into account the slope of the SQUID $I_{\mathrm{c}}(\Phi)$ modulation curve, we determined the strength of the qubit flux signal in the SQUID to be $\Phi_{\mathrm{sq,qb}} = 73$ mΦ_0. This is slightly higher than the expected design value of $\Phi_{\mathrm{sq,qb}} = M_{23} \cdot I_{\mathrm{c,qb}} = 53$ mΦ_0. Such an increase in the qubit signal strength might be due to a higher critical current of the qubit $I_{\mathrm{c,qb}}$ than designed, which goes along well with the fact that also the observed average qubit level splitting $\Delta_{01} \approx 13$ GHz was slightly higher than expected. An advantage of such distant SQUID peaks is that single-shot visibility is possible, i.e. a single measurement gives non-ambiguous information about the measured qubit state. However, the two peaks are much further apart than would be needed for this. As mentioned before in section 9.2.3, such a strong coupling might cause additional, unnecessary decoherence, so that the value of M_{23} should be reduced in future qubit designs.

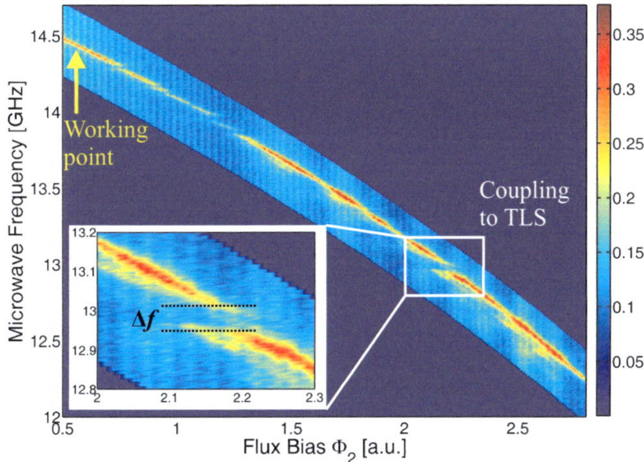

Figure 9.5: Spectrum of phase qubit Ch1. The measured P_{esc} is color coded. The inset shows a magnification of an avoided level crossing due to coupling of the qubit to a TLS. The working point for qubit characterization is highlighted in yellow.

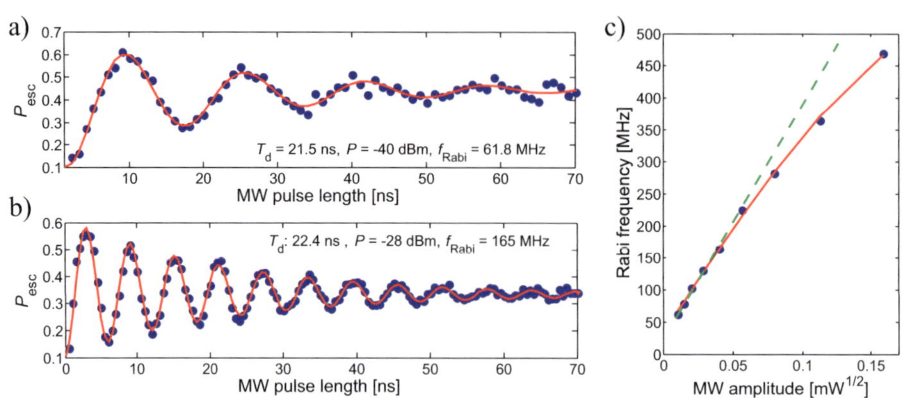

Figure 9.6: a) + b) Rabi oscillations of phase qubit Ch1 for different microwave powers (given at the source output). c) Rabi frequency versus applied microwave amplitude. For low MW amplitudes, the linear law of (9.6) is fulfilled (green dashed line), while for the entire range, the data can be explained by (9.7), as illustrated by the red line.

Figure 9.5 shows the spectrum of phase qubit Ch1. As expected, the qubit level splitting Δ_{01} changes with the applied flux bias Φ_2 and hence the potential tilting angle. At some points, avoided level crossings are present (see inset). These are due to the coupling of the phase qubit to some TLS in its environment. The coupling strength is given by the size of the avoided crossing Δf. This means that the qubit should exchange information with the TLS with a characteristic time $\tau_{\text{qb,TLS}} = 1/\Delta f$. For the most strongly coupled TLS, we found $\Delta f = 90$ MHz and hence $\tau_{\text{qb,TLS}} = 11$ ns. Although this is shorter than the observed qubit coherence time (see below), no coherent oscillation of information between the qubit and any of the found TLS could be observed.

For characterization of qubit Ch1, a working point far from any visible avoided level crossings was chosen. This was found at $\Delta_{01} = 14.48$ GHz and is marked in yellow in Figure 9.5. At first, Rabi oscillations were measured as described above for various applied MW powers. As can be seen in Figure 9.6, clear oscillations between the qubit states $|0\rangle$

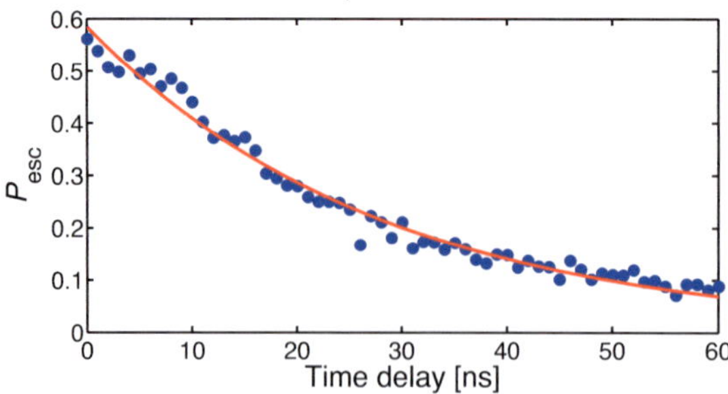

Figure 9.7: Relaxation time measurement on phase qubit Ch1. The obtained $T_1 = 25.9$ ns is longer than for similar Nb/AlO$_x$/Nb phase qubits measured by other groups (see section 9.5.3).

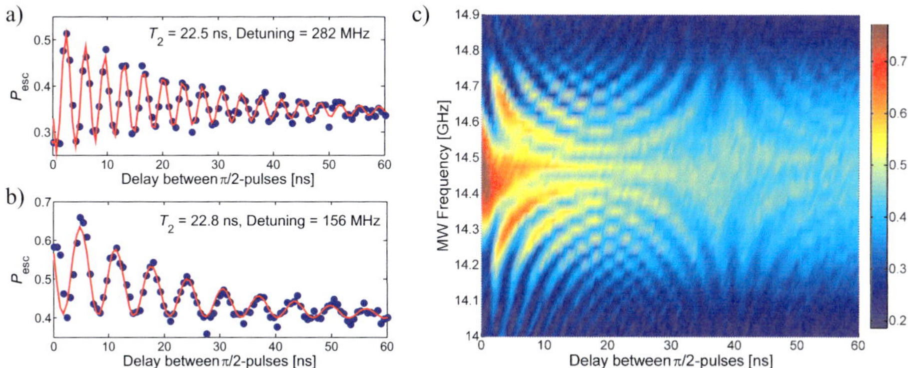

Figure 9.8: a) + b) Ramsey oscillations of phase qubit Ch1 for different MW frequencies f_{MW}. c) Surface plot of the escape probability (P_{esc}, color coded) as a function of f_{MW} and the delay time between the $\pi/2$-pulses.

and $|1\rangle$ with a characteristic decay time around $T_d \approx 22$ ns were obtained. Furthermore, Figure 9.6c shows that the dependence $f_{Rabi} \propto I_{MW}$ given by (9.6) is fulfilled for small MW amplitudes. Furthermore, it was tested whether the data could be explained in the entire amplitude range by including transitions to the second excited qubit state. An excellent agreement with theory was obtained, which can be seen by the red line in Figure 9.6c. It was obtained by a fit with $\Omega_R = \varkappa_1 I_{MW} \left[1 - \varkappa_2 (\varkappa_1 I_{MW})^2 \right] + \varkappa_3$ according to (9.7).

Measurement of the Rabi oscillations allowed calibration of the MW pulse length, so that now a controlled π-pulse could be applied in order to measure the relaxation time T_1. The result can be seen in Figure 9.7. The qubit indeed relaxes down to the ground state exponentially with a characteristic time $T_1 = 26$ ns. As will be discussed in more detail in section 9.5.3, this obtained relaxation time is longer than for any other reported Nb/AlO$_x$/Nb phase qubit of similar design.

To determine the dephasing time T_2, we performed Ramsey experiments as described in section 9.3.1. Figure 9.8 shows the obtained Ramsey oscillations for different applied MW frequencies f_{MW}. As can be seen in the surface plot in Figure 9.8c, for $f_{MW} = f_{01}$,

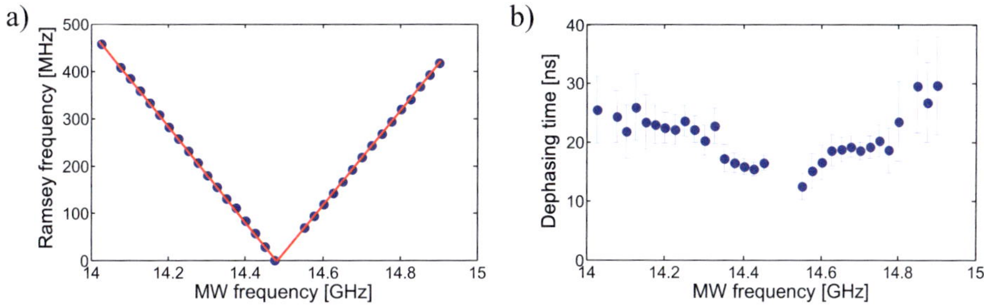

Figure 9.9: a) Ramsey frequency as a function MW frequency f_{MW}. b) Dephasing times T_2 and error bars for various f_{MW} values. The average value accounts for $T_2 = 21$ ns.

no oscillations occur and P_{esc} drops exponentially to $\approx \frac{1}{2}$ according to (9.9). In contrast, coherent oscillations were visible if f_{MW} was detuned from the Larmor frequency f_{01}. These were investigated in more detail by plotting the oscillation frequency f_{Ramsey} over the MW frequency, as shown in Figure 9.9a. As expected from (9.10), a linear behaviour $f_{\text{Ramsey}} \propto |f_{01} - f_{\text{MW}}|$ is found in nearly perfect agreement with theory. Figure 9.9b shows the determined dephasing times T_2 for various values of the driving frequency f_{MW}. The different T_2 are of the same order and an average value of $\langle T_2 \rangle = 21$ ns is found. Analogous to the measured relaxation time $T_1 = 26$ ns, this dephasing time is significantly longer than in any other reported Nb/AlO$_x$/Nb phase qubit of similar design (see section 9.5.3).

9.5.2 Characterization of Phase Qubit Ch2

Phase qubit Ch2, which should be identical to Ch1 except for the missing shunt capacitor, was characterized in the same run of the dilution refrigerator. Due to limitations of the measurement setup, no minimum was observed in the $I_c(H)$ curve of the read out DC SQUID. Consequently, the applied voltage in the bias coil could not be converted into applied flux and the slope of the SQUID modulation could not be determined. Hence, the coupling between qubit and SQUID $\Phi_{\text{sq,qb}}$ could not be calculated. Since the design values of M_{23} and $I_{\text{c,qb}}$ are identical to the ones of phase qubit Ch2 and the qubits were processed on the same chip, we can assume that the coupling will be similar to the one observed for Ch1.

The spectrum of phase qubit Ch2 can be seen in Figure 9.10. Again, avoided level crossings due to coupling to TLS are present. Unfortunately, the level splitting of the qubit lies around $\Delta_{01} \approx 23$ GHz, which is significantly higher than the expected value of $\Delta_{01} \approx 17.8$ GHz. Qubit spectroscopy revealed that the MW power arriving at the sample inside of the dilution refrigerator is strongly frequency dependent above 20 GHz and too strongly attenuated to even carry out qubit measurements above ≈ 24 GHz. This limited the possibilities of qubit measurement strongly.

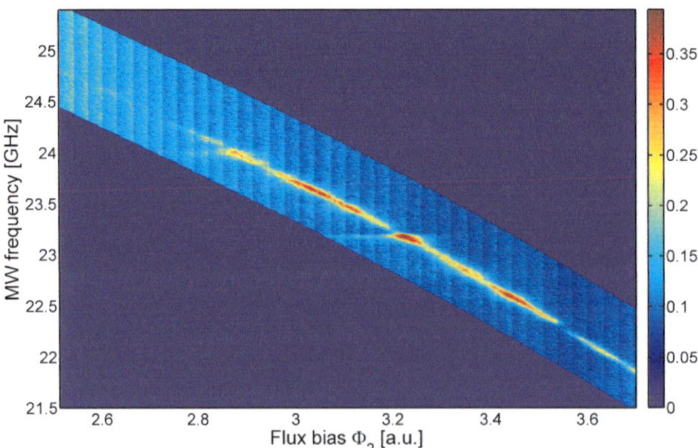

Figure 9.10: Spectrum of phase qubit Ch2. The measured P_{esc} is color coded. For higher frequencies, attenuation in the MW lines of the dilution refrigerator was so high that no sufficient MW amplitude was available at the sample for qubit spectroscopy.

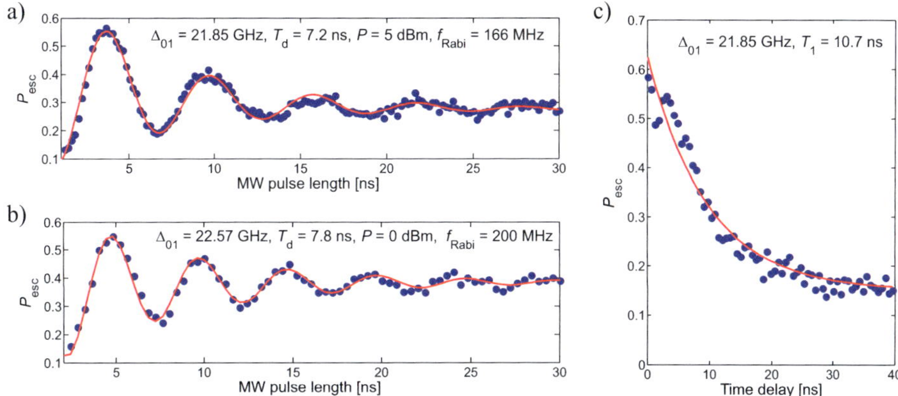

Figure 9.11: a) + b) Rabi oscillations of phase qubit Ch2 at different level splittings Δ_{01}. c) Relaxation time measurement for $\Delta_{01} = 21.85$ GHz.

We performed Rabi and T_1 measurements for various flux bias values Φ_2 and hence also for different qubit level splittings Δ_{01}. Exemplary measurements can be seen in Figure 9.11. For all chosen working points, we found coherent oscillations between the $|0\rangle$ and the $|1\rangle$ states with decay times around 7 ns. The corresponding relaxation times were about $T_1 = 10 - 11$ ns long. This is still longer than for many Nb/AlO$_x$/Nb phase qubits from commercial foundries (see section 9.5.3). Since the coherence times were significantly shorter than for qubit Ch1 and very large powers had to be applied in order to excite the qubit, no further measurements were carried out on phase qubit Ch2.

The reason for the stronger decoherence in qubit Ch2 than in sample Ch1 are discussed in the following. The only structural difference is the absence of a shunting capacitor in qubit Ch2. This means that most of the electrical field is stored in the tunneling oxide of the Josephson junction and not in the dielectric of the capacitor. However, decoherence in Ch2 could only be stronger if the losses in the AlO$_x$ barrier were larger than in the insulating SiO layer. Since it is difficult to investigate dielectric losses in the thermally grown, very thin tunneling oxide, it is impossible to say whether this is true. However, if the AlO$_x$ layer would limit the coherence times to $T_1 \approx 11$ ns, Al base phase qubits would probably not achieve coherence times of around 600 ns (see below). Consequently, the redistribution of the electrical field is probably not the correct explanation. In the PhD thesis of Jürgen Lisenfeld [120], it was also found that the coherence times in phase qubits with larger Δ_{01} values tended to be shorter. This is in agreement with our recent findings that the TLS density of states increases with rising frequency (for details see [Ska10] and [SKW$^+$11]). Hence, we conclude that the larger level splitting is responsible for the stronger decoherence in sample Ch2. Consequently, it should be taken care that Δ_{01} is not too high in future qubit designs.

9.5.3 Discussion of Possible Decoherence Mechanisms

While Al based phase qubits have reached coherence times up to 600 ns [92], for most Nb/Al-AlO$_x$/Nb phase qubits the relaxation times are shorter than $T_1 \simeq 10$ ns [14, 21, 40, 122] and even the longest observed times do not exceed 20 ns [22, 41, 123]. For all these devices, the dephasing times T_2 could either not be measured or were shorter than the T_1 times. (Only a changed phase qubit design, in which the qubit could be made insensitive to decoherence by operating it at an optimal point, resulted in longer coherence times about $T_1 \simeq 100$ ns for a Nb/Al-AlO$_x$/Nb based device recently [124].) However, a Nb/AlO$_x$/Nb process which is suitable for the fabrication of phase qubits would still be desirable, since this technology offers big advantages concerning scalability, integrability, reproducibility and yield. Since the coherence times observed in our phase qubit are longer than those in any other Nb/AlO$_x$/Nb phase qubits of similar design as mentioned above, we tried to learn some more about the reasons of decoherence in our devices.

First, we examined if the coherence times in our phase qubits could be limited due to sub-gap leakage. As discussed in section 5.4.2, this is not the case. Although the decoherence rate of $1/\tau_{qp} = 1/(100 \ \mu s)$ was only obtained by a rough estimation, it should definitely be smaller than the observed rate of $1/T_1 = 1/(26\,\text{ns})$. Also the ringdown times given by dissipation in the qubit should not play a role, as discussed in section 9.2.3, since they account for $T_1 = 0.7 - 2.2 \ \mu s$.

It follows that another decoherence mechanism must be present in our phase qubits. As already discussed in chapter 6, it was discovered recently that the coherence times in phase qubits were limited by the presence of spurious two-level systems (TLS) in their vicinity [23, 84]. The macroscopic manifestation of these TLS are dielectric losses [23, 84]. Since we could rule out other possible mechanisms of decoherence, we assume that the presence of such TLS is responsible for our limited coherence times, as well as for the decoherence in other Nb/AlO$_x$/Nb phase qubits. In the following, we discuss why we might have fewer TLS in our qubits and hence longer coherence times than in similar devices.

TLS have been found in the bulk dielectrics used for insulating layers in qubit fabrication [23] as well as on metal surfaces [95]. The dielectric losses in our Nb$_2$O$_5$ and SiO thin films are rather high (see chapter 6), and definitively higher than in SiN$_x$ (see Figure 6.5). The latter is the dielectric used for phase qubits fabricated by Hypres[2] or the PTB[3], which exhibited much shorter coherence times around 7 ns [125]. Hence, we do not believe that the losses in the insulating layers are responsible for our improved coherence times. Instead, the stronger decoherence in other phase qubits might either be due to higher subgap leakage or to TLS on the Nb surfaces. While the former hypothesis would be disallowed by the corresponding fabrication foundries, the latter is supported by the recent observation that the TLS responsible for the losses in superconducting coplanar waveguide (CPW) resonators stem from surface or interface states [88]. It seems likely that processes like RIE or CMP alter the Nb surface by ion bombardment, mechanical stress or chemical reaction. Since we have taken care to avoid such processes wherever possible (for example by using the Al hard mask technique, see section 4.3, or patterning the SiO insulation layer by lift-off), the enhanced coherence times in our phase qubits might indeed be due to the novel

---

[2]Hypres is a commercial foundry for Josephson junction circuits located in Elmsford, NY, USA
[3]Physikalisch-Technische Bundesanstalt, Braunschweig, Germany

fabrication process presented in chapter 4. This is supported by the fact that the quality factors of Nb based CPW resonators were found to be significantly larger if the Nb films were lifted-off instead of being patterned by reactive ion etching [96].

However, we still use RIE for trilayer and junction patterning. Consequently, we hope to further increase our T_1 and T_2 times by either improving or entirely omitting our employed RIE process. For example, fabricating phase qubits exclusively by lift-off processes and measuring the coherence times would yield information about the influence of RIE on the TLS density in Nb surfaces. If the coherence times do not significantly improve in this way, the decoherence might be due to TLS in native Nb oxides, which form as soon as Nb has contact with air. In this case, passivation of the Nb surfaces (for example by nitridization) before the wafers are taken out of vacuum might decrease the TLS density and result in longer coherence times. Additionally, the use of insulating layers with lower dielectric losses and hence fewer TLS should lead to longer coherence times. However, this will probably only yield a significant improvement after the problem with TLS in Nb surfaces is solved.

A more radical approach would be to entirely leave out amorphous dielectrics in the fabrication process. This would be possible by realizing the Josephson junctions as weak links, such as Dayem bridges [25] or other nanobridges. Additionally, such systems could be realized in a one-layer process and would hence be faster in production. However, the conditions for observation of the Josephson effect with sinusoidal current-phase relationship according to (2.7) and phase coherent supercurrent flow, namely that the characteristic bridge dimensions are smaller than the coherence lengths $\xi(T)$ and ξ_0 (see section 2.2.1) are very difficult to fulfill. For example, $\xi(T) = 18$ nm has been determined for Nb at 4.2 K in section 5.2. First experiments on the realization of the Josephson effect in Nb Dayem bridges at the IMS have been carried out in the *Studienarbeit* of Max Bauer [Bau10]. However, the Josephson effect could not yet be shown in the fabricated structures and further investigations would be necessary.

9.6 Conclusions

The fabrication process for high quality sub-μm to μm-size Josephson junctions developed in this thesis was successfully used to fabricate superconducting phase qubits. The qubit level splittings were higher than expected, which led to experimental difficulties in the investigation of the capacitively unshunted qubit Ch2. However, relaxation times around 10-11 ns could be measured for this sample, which is longer than for many other Nb based phase qubits. For the capacitively shunted qubit Ch1, a relaxation time of $T_1 = 26$ ns and a dephasing time of $T_2 = 21$ ns were obtained. This is longer than for any other reported Nb/AlO$_x$/Nb phase qubit of similar design. Rabi and Ramsey type experiments were performed and show excellent agreement with theory. Finally, possible decoherence mechanisms in our qubits were discussed. The Al hard mask technique developed in this thesis might be a reason for the long coherence times, since it helps to avoid chemical processes like RIE or CMP.

10 Summary

In this thesis, a technological process for the fabrication of high-quality Nb/Al-AlO$_x$/Nb Josephson Junctions with sizes down to the sub-μm regime has been developed and successfully used for various macroscopic quantum experiments.

The basis for such experiments is the fact that the charge transport in superconductors is carried out by Cooper pairs, which have zero spin and hence exhibit boson-like behavior. This is why the whole superconductor can be described by only one macroscopic quantum mechanical wavefunction. If this superconductor is interrupted by a thin insulating barrier, the macroscopic wavefunction exhibits a well-defined phase difference, which is connected to the current and voltage over this element. This is called the Josephson effect and such a device is known as a Josephson junction. The dynamics of such a system can be described by a tilted washboard potential containing a macroscopic phase particle, which can escape from a potential well by quantum tunneling. Consequently, Josephson junctions are ideal systems for the investigation of macroscopic quantum effects and the fabrication of macroscopic quantum devices such as quantum bits.

One possible experimental realization of such elements is given by the Nb/Al-AlO$_x$/Nb technology, which has been the standard process for many applications over two decades and exhibits very high yield, reproducibility, scalability and low parameter spread.

Despite the many advantages of Nb/Al-AlO$_x$/Nb technology for classic applications such as RFSQ and voltage standards, the usage of this technology for high-performance quantum bits remains a challenge. Superconducting phase qubits fabricated by foundries specialized on high yield processes have shown very poor coherence times below 10 ns. Consequently, the focus in this thesis was put on the development of a fabrication process especially suited for quantum devices and experiments.

This was done in two steps. At first, the standard fabrication process involving photolithography was optimized. Through a careful analysis of the junction quality before the start of this thesis, processing steps which had to be improved where identified. A working point for minimal Nb film stress was found by measuring the curvature of the silicon substrate wafers. The Al thickness was changed to the value of 7 nm, which was found to be ideal in literature. The reactive ion etching (RIE) process was optimized considering slow, precise, uniform and clean etching with the use of response surface methodology (RSM). Finally, the anodic oxidation of the lower electrode sidewalls was introduced by changing the order of the processing steps.

Characteristic measurements on Josephson junctions fabricated with the optimized process revealed a clearly improved quality in comparison to before the start of this thesis. Furthermore, the obtained quality parameters $V_{\mathrm{gap}} > 2.8$ mV, $R_{\mathrm{sg}}/R_{\mathrm{N}} \gtrsim 30$, $I_{\mathrm{c}}R_{\mathrm{N}} \approx 1.8$ mV and $V_{\mathrm{m}} \gtrsim 60$ mV are excellent. The yield was found to be larger than 98 %, which is largely sufficient for all aspired quantum experiments.

The second step in order to obtain a fabrication process for high-performance quantum devices was to develop a technology for sub-µm to µm-size Josephson junctions. Here, the novel concept of an Al hard mask was used for junction definition. The hard mask is created by positive e-beam lithography, Al sputtering, and lift-off. It is oxidized during the anodization, protects the underlying Nb surface on top of the Josephson junction and can be removed by a clean KOH wet etching process. In order to obtained small vias in the SiO insulation layer, the employed e-beam lithography process and the SiO thermal evaporation were carefully optimized. Before the Nb wiring layer is deposited, an *in-situ* pre-cleaning by low-energy Ar ion beam etching was introduced, so that an electrical contact through the narrow vias is obtained.

Josephson junctions were fabricated with this newly developed process down to areas of $0.5\ \mu m^2$. Their characterization showed that their quality was just as excellent as that of the junctions fabricated with the optimized standard process. In addition, detailed measurements of the subgap regime were carried out with a voltage bias setup at mK temperatures. These investigations showed that the RCSJ model describes the junctions with high accuracy and that the coherence times in qubits fabricated with this technology should not be limited by quasiparticle conductance.

One of the main decoherence mechanisms in superconducting phase qubits was found to be coupling of the qubit to spurious two-level systems (TLS) in its environment. Since the macroscopic manifestation of such TLS are dielectric losses, the development of a reliable method for the determination of dielectric losses is an important part of the development of a fabrication technology for quantum devices. In this thesis, a reliable method for direct measurements of dielectric losses in thin films employing superconducting resonators was developed. This method needs no input of any fitting parameters and allows to measure the losses in the film volume as well as the metal/dielectric interfaces quantitatively. The losses in thin films of Nb_2O_5, SiO, SiO_2, SiN_x and a-Si:H were measured at cryogenic temperatures and GHz frequencies, showing that the losses in the volume of such amorphous materials exceed the ones on the interfaces by far. Furthermore, the frequency dependences of the losses under the above conditions were investigated for the first time. The corresponding results suggest that the TLS exhibit many-body interactions, which is an important fact for their theoretical modeling. The investigation of the dielectric losses in multi-layers showed a good agreement between measurement and theoretical expectation, namely that the layers contribute to the total losses weighted with their permittivities. This is especially important for Josephson junction fabrication, since the Nb_2O_5 layer, which is crucial for junctions with low subgap leakage, exhibits high losses, but also has a very high permittivity. Consequently, it should still be possible to use it in qubit fabrication as long as it is combined with a layer having lower losses and lower permittivity.

The newly developed technological process for the fabrication of high-quality sub-µm to µm-size $Nb/Al\text{-}AlO_x/Nb$ Josephson junctions was used for three macroscopic quantum experiments. At first, the classic proof that a macroscopic quantum effect is observed in Josephson junctions, namely the observation of tunneling of a phase particle out of a potential well, was aspired. Such a macroscopic quantum tunneling (MQT) experiment was carried out systematically for varying junction size and varying magnetic field, which both had not been investigated experimentally before. In order to perform the experiments in the low damping limit, the damping at frequencies relevant for MQT was significantly decreased by using decoupling inductors. The crossover from the thermal to the quantum regime was found to have a clear and systematic dependence on junction size, which was in perfect agreement with theory. Furthermore, the dependence of the escape temperature with applied magnetic flux agreed excellently with theory. Additional structures in the corresponding measurement results might be due to the occurrence of Fiske resonances in the junctions, which will be analyzed in collaboration with theory groups in the future.

Since the MQT experiments confirmed that the developed Josephson junction technology is excellently suited for quantum experiments, the dynamics of a more complex structure was investigated, namely in an *LC* shunted Josephson junction. In such a system, two phase drops occur in the system, so that it has to be described by a two-dimensional potential. This leads to the occurrence of two new energy scales (instead of the single junction energy scale) in the system. The circuit elements were simulated and designed carefully, in order to investigate the dynamics for different sets of parameters. MQT measurements performed on these samples showed in principle the expected behavior, but will have to be analyzed in collaboration with theory groups in the future to achieve a detailed understanding. By probing the energy levels in the potential well with microwave spectroscopy, it was shown that such a system indeed acts as one single quantum system. This is remarkable since the circuit has a size of $200 \times 650 \ \mu m^2$, which is larger than for most superconducting qubits. Furthermore, an excellent agreement between the theoretically expected behavior, calculated using the design parameters, and the measurement data was found. These results are crucial for the understanding of the dynamics in high-temperature Josephson junctions, where such shunting elements cannot always be avoided in fabrication. Furthermore, the findings are important for superconducting quantum circuits such as quantum bits, since such systems are often shunted capacitively or inductively at will. Consequently, the fact that there is indeed a second energy scale in an *LC* shunted Josephson junction should be considered in quantum device design and operation.

For the final macroscopic quantum experiment in this thesis, the new fabrication process was used to build superconducting phase qubits, which consist of a Josephson junction embedded in a superconducting ring. Such a structure is described by a double-well potential, where each minimum corresponds to a circulating current in opposite direction. The two lowest energy levels in one well are used as the qubit states, and read out is performed by letting the excited state tunnel to the other potential minimum. The corresponding change in generated magnetic flux is detected by a DC SQUID.

The qubit circuits were designed carefully by simulating all involved inductances and calculating the eigenenergies in the qubit potential well numerically. Two qubits were investigated, of which one was capacitively shunted to lower the microwave frequency required for operation. The experiment showed that the qubit level splittings were higher than expected, which led to experimental difficulties in the investigation of the capacitively unshunted qubit Ch2. However, relaxation times around 10-11 ns could be measured for this sample, which is still longer than for many other Nb based phase qubits. For the capacitively shunted qubit Ch1, a relaxation time of $T_1 = 26$ ns and a dephasing time of $T_2 = 21$ ns were obtained. This is longer than for any other reported $Nb/AlO_x/Nb$ phase qubit of similar design. By coherent manipulation of the qubit states with microwaves, Rabi and Ramsey type experiments were performed and show excellent agreement with theory. Finally, possible decoherence mechanisms in the qubits were discussed. Since subgap leakage and losses in bulk dielectrics could be excluded as reasons for the long coherence times in the devices, it can be assumed that the decoherence is mainly due to TLS on the Nb surfaces. Consequently, the Al hard mask technique developed in this thesis might be a reason for the longer coherence times, since it helps to avoid chemical processes like RIE or CMP, which might be responsible for the TLS formation on metal surfaces.

In this thesis, a new technological process for the fabrication of sub-μm to μm-size Nb/Al-AlO_x/Nb Josephson junctions was developed, which led to results in excellent agreement with theory for all three performed macroscopic quantum experiments. Some of the observed effects were unexpected and have to be investigated in collaboration with theory groups in the future. For superconducting phase qubits, the obtained coherence times were longer than for any other reported Nb based phase qubit of similar design. Consequently, this process can be used in the future for many macroscopic quantum applications. Further optimization of the SiO via definition, possibly implying a two-layered e-beam resist, might allow the fabrication of even smaller junctions, so that also the fabrication of flux qubits might be possible. In order to obtain even longer coherence times in Nb/Al-AlO_x/Nb phase qubits, fabrication of the devices exclusively by lift-off processes is advisable, since this entirely omits chemical processes like RIE. As a next step, insulating layers with lower dielectric losses than SiO should be used. A method to quantitatively measure these losses in the qubit working regime and a Josephson junction like environment was developed within this thesis.

A Magnetic Flux Quantization

Since all Cooper pairs can be described by a single wavefunction, one can expect macroscopic quantum phenomena in superconducting systems. In 1950, Fritz London predicted that the magnetic flux Φ penetrating a superconducting ring should be quantized, which was experimentally confirmed by R. Doll and M. Näbauer eleven years later. As this flux quantization plays an important role for the principles of SQUIDs and phase qubits, it will be discussed here in some detail.

In general, when a magnetic flux penetrates a superconducting ring, a circulating current will be induced. Ginzburg and Landau showed in 1950 in their macroscopic, phenomenological theory that the current density in a superconductor being in a magnetic field $\vec{B}(\vec{x})$ having a vector potential $\vec{A}(\vec{x})$ is given as [24]

$$\vec{j}_s(\vec{x}) = \frac{q_s \hbar}{2 m_s i} \left[\psi^*(\vec{x}) \cdot \vec{\nabla} \psi(\vec{x}) - \psi(\vec{x}) \cdot \vec{\nabla} \psi^*(\vec{x}) \right] - \frac{q_s^2}{m_s} \vec{A}(\vec{x}) \psi^*(\vec{x}) \psi(\vec{x}) . \tag{A.1}$$

Here, $\hbar = h/(2\pi)$ is the reduced Planck constant and ψ^* denotes the complex conjugate of ψ. In a homogeneous superconductor, the Cooper pair density n_s is constant so that only the phase ϕ in (2.5) depends on space. This turns (A.1) into

$$\mu_0 \lambda_L^2 \vec{j}_s(\vec{x}) = \frac{\hbar}{q_s} \vec{\nabla} \phi(\vec{x}) - \vec{A} . \tag{A.2}$$

If we now integrate along the ring and take a path $\vec{dl}$ far enough inside the ring (compared with λ_L) so that no currents flow, the left side of (A.2) gives zero. For the first term on the right side, we should obtain

$$\oint \vec{\nabla} \phi(\vec{x}) \cdot \vec{dl} = 2\pi n , \tag{A.3}$$

with n being an integer, since ψ must be continuous. Using Stokes' theorem and $\vec{\nabla} \times \vec{A} = \vec{B}$, the second term on the right side of (A.2) can be transformed into

$$\oint \vec{A}(\vec{x}) \cdot \vec{dl} = \int_S \vec{\nabla} \times \vec{A}(\vec{x}) \cdot d\vec{S} = \int_S \vec{B}(\vec{x}) \cdot d\vec{S} = \Phi , \tag{A.4}$$

where S is the area penetrated by the magnetic flux Φ and surrounded by the superconducting ring. Finally, we see that (A.2) turns into

$$\Phi = n \cdot \frac{h}{q_s} = n \cdot \Phi_0 . \tag{A.5}$$

This shows that the magnetic flux is quantized with the magnetic flux quantum Φ_0. Knowing that the superconducting charge carriers are Cooper pairs, we assume that $q_s = 2e$ and hence $\Phi_0 = h/(2e) = 2.0678 \cdot 10^{-15}\,\text{Tm}^2$. This value was also found experimentally, which impressively confirms the existence of Cooper pairs. If the externally applied magnetic field does not correspond to $n\Phi_0$, circulating currents in the superconducting ring arise, so that the total magnetic flux is still compensated to a multiple of the flux quantum.

B Detailed Processing Parameters

B.1 Process Before the Start of This Thesis

In the following list, details of the processing parameters as used before the start of this thesis are given (the names of the layers are given in brackets). Potential flaws in this process as well as its optimization are discussed in chapter 4.

Deposition of the trilayer in the UTS 500 vacuum system:

- Pre-cleaning of the wafer by an RF Ar plasma in the load lock.
- DC magnetron sputtering of Nb for 240 s (nominal thickness: 200 nm) at a power of 300 W and an Ar working pressure of 5.5 mTorr in the main chamber.
- DC magnetron sputtering of Al for 15 s (nominal thickness: 7 nm) at a power of 100 W and an Ar working pressure of 5.4 mTorr in the main chamber.
- Oxidation of the Al layer for 30 min at room temperature in the load lock in a pure oxygen atmosphere at a pressure depending on the required critical current density j_c.
- DC magnetron sputtering of Nb for 120 s (nominal thickness: 100 nm) in the main chamber with the same parameters as for the bottom electrode.

Definition of the Josephson junctions (J):

- Positive photolithography
- Reactive-ion-etching (RIE) of top trilayer electrode (M2b) at a working pressure of 260 mTorr, CF_4 flow of 29 sccm and O_2 flow of 5.9 sccm.
- Anodic oxidation at a voltage of 30 V in an aqueous solution of $(NH_4)B_5O_8$ and $C_2H_6O_2$ at room temperature.

Definition of the bottom electrode (M2a):

- Positive photolithography
- RIE (same parameters as described above) to remove Nb_2O_5 created during anodic oxidation.
- Ion beam etching (IBE) through the Al-AlO$_x$ layer.
- RIE (same parameters as described above) to pattern the lower trilayer electrode (M2a)

Definition of the insulation layer (I):

- Negative photolithography
- Thermal evaporation of a 300 nm thick SiO layer in two deposition steps of 150 nm each.
- Lift-off

Definition of the wiring layer (M3):

- Negative photolithography

- Pre-cleaning of the wafer by an RF Ar plasma in the load lock of the UTS 500 system.
- DC magnetron sputtering of Nb for 420 s (nominal thickness: 350 nm) in the main chamber of the UTS 500 system (same parameters as described above).
- Lift-off

B.2 Technological Process as Developed During This Thesis

In chapter 4, the optimization of certain processing steps is discussed in detail. Furthermore, an overview of the process and the order of the single steps are given. In this section, parameters of the single employed processes are discussed in detail.

Photolithography

The parameters of the photolithography varied over time and were optimized whenever it was found necessary. The most prominent change was replacing the developer for the positive lithography, since the use of the AZ Developer did not fully develop the resist next to the defined structures anymore. Before the actual photolithography, the thicker edge of the resist was removed by 42 s exposure and 30 s development. With negative photolithography, the inverse of the mask structure is patterned into the resist.

Table B.1: Parameters of negative and positive photolithography. MIF stands for metal-ion free.

	Positive photolithography	Negative photolithography
Resist	AZ 5214E	AZ 5214E
Spinning	6000 rpm for 60 s	6000 rpm for 60 s
Baking	4 min at 85 °C	4 min at 85 °C
Exposure	10-11 s	5 s
Post-baking	-	5 min at 120 °C
Flood exposure	-	1 min
Developer	AZ Developer, later AZ 726 MIF	AR 300-47 MIF
Development	around 1 min	around 1 min 40 s

Deposition of Trilayer

The trilayer deposition took place in the UTS 500 vacuum system. It contains a load lock and a main chamber. In the following, the detailed process parameters are given.

- Pump load lock to base pressure $< 10^{-5}$ mbar. RF pre-cleaning in Ar plasma at a pressure of $p = 4 \cdot 10^{-2}$ mbar and a RF power of 30 VA for 3 min.
- Pump load lock and bring wafers to main chamber having a base pressure $p \leq 5 \cdot 10^{-7}$ mbar. Set Ar flowmeter to 2.0 V and rotate butterfly valve above turbopump until $p = 7.2$ mTorr $(= 9.6 \cdot 10^{-3}$ mbar). Sputtering of Nb takes place at a power of $P = 300$ W with 3 min pre-sputtering.

- Reset the butterfly valve to obtain $p = 5.4$ mTorr ($= 7.2 \cdot 10^{-3}$ mbar). Sputtering of Al takes place at a power of $P = 100$ W with 10 min pre-sputtering. The same parameters are used for deposition of the Al hard mask.
- Pump main chamber. The load lock is rinsed three time with oxygen and pumped again before the wafer is brought into it. The desired oxidation pressure is set and the oxidation is (usually) performed for 30 min at room temperature. For higher pressures $p_{oxy} \geq 1$ mbar, the turbopump is separated from the load lock and the oxidation is performed with a static pressure. For lower pressures $p_{oxy} \leq 1$ mbar, the oxidation is performed in a steady state with constant Ar inflow and constant pumping power.
- After pumping of the load lock, the wafer is transferred to the main chamber again. Nb is sputtered with the same parameters as before.

Reactive Ion Etching (RIE)

Optimization of the RIE process is discussed in section 4.2.2 with detailed modeling parameters given in Appendix C. In the following, the details of an etching procedure are given:

- Put sample in load lock and pump it.
- The main chamber is preconditioned by creating a plasma with the standard etching parameters for 6 min.
- Bring the sample into the main chamber if pressure in load lock is below $p \approx 6 \cdot 10^{-5}$ mbar.
- The etching takes places at an CF_4 flow of 49 sccm, an O_2 flow of 21 sccm, a pressure of 200 mTorr (=0.27 mbar) and an RF power of 100 W. Using the matchbox of the system, the reflected power is kept below 3 W. Usually, for 100 nm of Nb, RIE is carried out for 1 min 30 s.

Ion-Beam Etching (IBE)

The IBE parameters were taken over from before the start of this thesis. The only purpose of the IBE is to remove the Al/AlO$_x$ tunneling barrier, which cannot be etched by RIE.

- Put sample in load lock and pump it.
- The plasma and later the ion beam are already started in the main chamber in order to stabilize it.
- Plasma and ion beam are turned off. The sample is brought into the main chamber with an elevator.
- The plasma is started in the main chamber at an Ar flow of 1.5×4 sccm, a pressure of $p \approx 2.5 \cdot 10^{-4}$ mbar and an RF power of about 100 W.
- The positive grid voltage of 250 V and the negative grid voltage of 200 V are applied, so that the ion beam is started. The RF power reset so that a positive grid current of ≈ 48 mA is reached.
- The tilted, rotation sample table is turned on. The shutter above the wafer is opened, so that etching begins. It takes place for 2×5 min with a pause of 5 min in which the shutter is closed.

Electron-Beam Lithography

Electron-beam lithography was used to define small Josephson junctions and small vias through the SiO insulating layer. Especially for the latter, a thorough optimization of the e-beam parameters was necessary. In Table B.2, the employed parameters are given.

Table B.2: Parameters of the employed e-beam lithography. The resist thickness was measured using the TENCOR profilometer.

	JJ definition	SiO vias
Resist	PMMA 950 3%	PMMA 950 5%
Spinning	8500 rpm for 60 s	8500 rpm for 60 s
Baking	5 min at 175 °C	5 min at 175 °C
Thickness	235 nm	465 nm
High voltage	9 kV	9 kV
Aperture	10 µm	10 µm
Dose	50 µC/cm^2	63 µC/cm^2
Development	50 s in AR 600.56	50 s in AR 600.56

Anodic Oxidation

Quite a few details of the anodic oxidation are already discussed in section 4.2.3. The voltage for anodic oxidation can be chosen freely, but was mostly taken to be 20 V to 30 V. The current limitation on the voltage source is set to minimum in order to obtain a slow Nb_2O_5 growth. After the voltage is applied, it takes about 20-30 s to reach the set value. The current during this time usually accounts for 0.68 mA. Anodization usually takes place for 6 min and a final current of 2-3 µA is normally reached.

Wet Etching of the Al Hard Mask

The Al hard mask is etched in 20 wt% KOH, which is diluted 1:1 with H_2O. The etching progress can be controlled by eye. Usually, the wafers are kept in the KOH for about 4 min 30 s to make sure that no remains of the hard mask are left.

Thermal evaporation of SiO

After the samples are built into the vacuum chamber, the latter is usually pumped over night to reach a base pressure $2 - 7 \cdot 10^{-7}$ mbar. Then, the current through the crucible is carefully increased, so that the pressure in the chamber rises. This bake-out is continued until the pressure starts dropping again. Evaporation takes place in four steps. For each step, a base pressure $< 5 \cdot 10^{-7}$ mbar and a working pressure of $< 8 \cdot 10^{-7}$ mbar should be reached. The film thickness is measured *in-situ* with an INFICON oscillating crystal. The current through the crucible is set to 5.4 - 5.6 A, so that a deposition rate of 0.5 - 0.8 nm/s is reached.

Deposition of the M3 wiring layer

The deposition of the Nb wiring layer takes place in the UNIVEX 450 vacuum system. As discussed in section 4.3.3, pre-cleaning before the sputtering is important especially for small SiO vias. In the following, the details of the M3 deposition are given:

- Put sample in load lock and pump it to below $5 \cdot 10^{-6}$ mbar.
- After the sample is transferred to the main chamber (having a base pressure $< 2 \cdot 10^{-7}$ mbar), pre-cleaning takes places with an ion gun. This is done at an Ar flow of 5 sccm, an Ar pressure of $1.3 \cdot 10^{-3}$ mbar, a beam voltage of 100 V and a beam current of 10 mA for 10 min with a rotating sample holder.
- Nb is sputtered on the rotating sample holder at a sputter current of 175 mA and a pressure of $5.0 \cdot 10^{-3}$ mbar. For thick films > 300 nm, sputtering should be carried out in two steps in order to avoid heat-induced damage to the photoresist. For each deposition run, pre-sputtering is performed for 2 min by sweeping through the sputtering current from 40 mA to 400 mA multiple times.

C Response Surface Methodology Modeling of the RIE Process

Since plasma processes like RIE cannot be modeled analytically, the etching properties at a certain set of parameters cannot be extrapolated from the properties at another set of parameters. Consequently, the parameter space has to be searched experimentally for a working point with the required etching properties. This search can be carried out systematically and efficiently with the response surface methodology (RSM) [126, 127]. This is an empirical modeling tool which allows to significantly reduce the required experimental points. Since plasma processes can exhibit a non-monotonous dependence of the response to certain parameters, a modeling formula at least up to quadratic terms is required. This means that for a process with three independent parameters, a classic modeling would require $3^3 = 27$ data points (see Figure C.1a). With RSM and a Box-Behnken design (see Figure C.1b), the number of required points can be reduced to 13. One point (preferably in the center) can be measured several times in order to determine the process variance.

In the modeling of the RIE process, the power was held constant at $P = 100$ W and the three parameters pressure p, total flow $F = F_1 + F_2$ and oxygen content $O = F_2/F$ were varied. The evaluated parameter space can be found in Table C.1. The center point was replicated five times, so that 17 measurements were carried out. For data point i, the response Y_i (e.g. the etching rate of Nb $r_{e,\mathrm{Nb}}$) is given by

$$Y_i = d_0 z_{i,0} + d_1 z_{i,1} + d_2 z_{i,2} + d_3 z_{i,3} + d_4 z_{i,4} + d_5 z_{i,5} + d_6 z_{i,6} + d_7 z_{i,7} + d_8 z_{i,8} + d_9 z_{i,9} . \quad \text{(C.1)}$$

Here, the $z_{i,j}$ are combinations of the parameter values and the d_j are the coefficients, which are equal for all data points. In detail, $z_{i,0} = 1$, $z_{i,1} = p_i$, $z_{i,2} = O_i$, $z_{i,3} = F_i$, $z_{i,4} =$

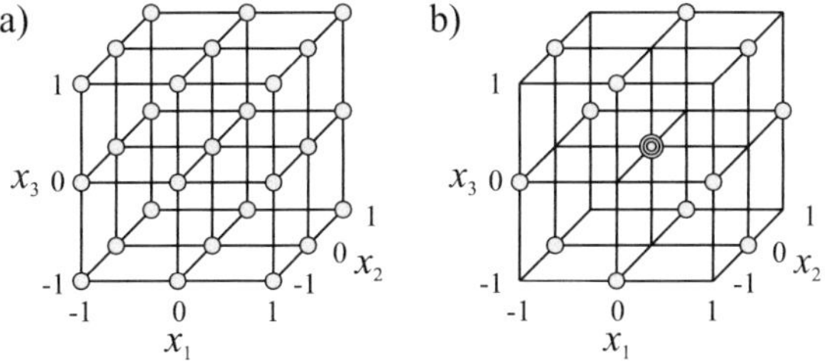

Figure C.1: a) A classic approach would require 27 data points in order to model a process with 3 parameters up to the second order. b) With the Box-Behnken design, the number of required points is reduced to 13 points. The center point can be measured several times in order to determine the variance of the process.

$p_i \cdot O_i$, $z_{i,5} = p_i \cdot F_i$, $z_{i,6} = O_i \cdot F_i$, $z_{i,7} = p_i^2$, $z_{i,8} = O_i^2$ and $z_{i,9} = F_i^2$. The mixed terms ensure that also interactions between different parameters are taken into account. Combining all 17 equations of type (C.1) into one results in the matrix equation

$$\begin{pmatrix} Y_1 \\ Y_2 \\ \vdots \\ Y_{17} \end{pmatrix} = \begin{pmatrix} z_{1,0} & z_{1,1} & \cdots & z_{1,9} \\ z_{2,0} & z_{2,1} & \cdots & z_{2,9} \\ \vdots & & \ddots & \vdots \\ z_{17,0} & z_{17,1} & \cdots & z_{17,9} \end{pmatrix} \cdot \begin{pmatrix} d_0 \\ d_1 \\ \vdots \\ d_9 \end{pmatrix}. \tag{C.2}$$

Since the vector $\vec{Y}$ and the matrix z are simply given by the experimental parameters, it is possible to calculate the coefficient vector $\vec{d}$ according to

$$\vec{d} = (z^T \cdot z)^{-1} \cdot z^T \cdot \vec{Y}. \tag{C.3}$$

Table C.1: Parameter space for RSM modeling of the RIE process.

Parameter	RSM value -1	RSM value 0	RSM value 1
Pressure p [mTorr]	100	150	200
Oxygen content O [%]	10	30	50
Total Flow F [sccm]	10	40	70

Results

The RSM modeling was carried out in the Box-Behnken design (see Figure C.1b) for the Nb etching rate $r_{e,Nb}$ and the resist etching rate $r_{e,res}$; the measurement data can be found in Table C.2. They were obtained by taking a Nb film, usually of 600 nm thickness, creating a resist mask on top of it and etching a certain thickness < 600 nm. By measuring the profile after lithography, after etching and after resist removal, the etching rates for Nb and the photoresist could be determined.

The parameter space given in Table C.1 proved to be too large for a linear modeling $Y = r_{e,Nb}$. The obtained etching rates varied over several orders of magnitude, so that the obtained model returned negative $r_{e,Nb}$ and $r_{e,res}$ values for some parts of the parameter space. Consequently, a logarithmic scaling $Y = \ln(r_{e,Nb})$ and $Y = \ln(r_{e,res})$ was used and gave much better results. For six new points within the parameter space, mean relative deviations from the fit curves were calculated. These accounted for 27 % for the Nb etching rate and 20 % for the selectivity $r_{e,Nb}/r_{e,res}$. In order to improve the precision of the modeling, these 6 points were included in the RSM model, so that 23 data points were evaluated. Now, for six randomly chosen values out of these 23 points, a mean relative deviation of 10 % for $r_{e,Nb}$ and 4 % for the selectivity was calculated. This means that the etching rates can be reliably predicted for any point within the parameter space. Since the latter is rather large and the etching rates vary strongly, this is an excellent result. The determined modeling coefficients d_j are given in Table C.3. More details about the foundations of plasma etching as well as this particular RIE modeling procedure can be found in the *Studienarbeit* of Maher Rezem [Rez11].

Table C.2: Experimental data points for RSM modeling of the RIE process. Points 18 to 23 were used as test points for the Box-Behnken modeling with 17 points and as further data points for the second modeling.

Data point	Pressure p [mTorr]	O_2 content O [%]	Total flow F [sccm]	$r_{e,Nb}$ [nm/min]	$r_{e,res}$ [nm/min]
1	150	30	40	110	87
2	150	30	40	89	79
3	150	30	40	111	76
4	150	30	40	100	67
5	150	30	40	110	67
6	150	10	10	808	1
7	200	30	10	430	42
8	150	50	10	121	123
9	100	30	10	481	71
10	200	10	40	560	2
11	200	50	40	71	218
12	100	50	40	41	271
13	100	10	40	561	34
14	150	10	70	483	18
15	200	30	70	85	69
16	150	50	70	35	275
17	100	30	70	44	123
18	200	17.1	35	618	1
19	100	17.1	35	230	36
20	150	17.1	35	318	1
21	100	29.2	21.2	127	78
22	100	29.2	41	60	110
23	100	29.2	66.4	53	126

Table C.3: Resulting coefficients from the RSM modeling. The physical units of the coefficients can be deduced knowing the units given in Table C.1 and that $[r_{e,Nb}]$ =nm/min.

Coefficient	$\ln(r_{e,Nb})$, 17 points	$\ln(r_{e,Nb}/r_{e,res})$, 17 points	$\ln(r_{e,Nb})$, 23 points	$\ln(r_{e,Nb}/r_{e,res})$, 23 points
d_0	4.6409	0.3114	4.6191	0.7013
d_1	0.1367	0.6560	0.2864	0.9039
d_2	-1.1506	-2.9388	-1.1808	-3.2377
d_3	-0.7210	-1.3136	-0.7108	-1.3337
d_4	0.1377	-0.5162	0.0470	-0.7991
d_5	0.1926	0.2059	0.1333	0.0185
d_6	-0.1815	0.3400	-0.1718	0.4354
d_7	0.1639	-0.0380	0.1048	-0.1157
d_8	0.3537	1.0836	0.4266	1.0846
d_9	0.3119	0.5816	0.3118	0.2646

D Calculation of the Energy Levels for a Phase Qubit

As has been explained in section 9.2.1, knowledge of the qubit level splitting Δ_{01} and the anharmonicity Δ_{12} is important for qubit design. Consequently, the energy eigenstates E_n in the potential well need to be calculated. This is done by solving the stationary Schrödinger equation

$$\hat{H}\,|\psi\rangle = E\,|\psi\rangle \tag{D.1}$$

with the Hamiltonian

$$\hat{H} = \frac{\hat{p}^2}{2M} + U(\hat{\varphi})\,. \tag{D.2}$$

Here, the potential $U(\varphi)$ is given by (2.36) and the mass of the Josephson phase particle is given by $M = C(\Phi_0/2\pi)^2$, as already discussed for the RCSJ model in section 2.2.2. The momentum operator is given by

$$\hat{p} = -i\hbar\frac{\partial}{\partial\varphi}\,. \tag{D.3}$$

Equation (D.1) is conveniently solved by choosing an appropriate basis, filling the matrix $\hat{H}$ and then diagonalizing it. Since for a phase qubit, the phase operator $\hat{\varphi}$ is the well defined variable (and not the canonically conjugate variable charge $\hat{q}$, as e.g. for a charge qubit), it is apparent to work in phase space. Since only bound states in the potential well shall be calculated, the φ space is restricted to the corresponding part of the potential. Furthermore, the continuous variable φ has to be discretized in order to fill the matrix. For this, a dimension N of the Hilbert space is chosen, so that values φ_k with $k = 1\ldots N$ are obtained. The larger the value of N, the more precise is the calculation, but the longer is the calculation time.

After the matrix elements were calculated analytically (see below), a *Matlab*[1] tool was programmed, which is able to calculate the eigenvalues of $\hat{H}$ numerically. This tool was then used in sample design in order to obtain the desired energy separation of the two qubit levels $|0\rangle$ and $|1\rangle$. It was found that for a Hilbert space dimension larger than $N \approx 1000$, no significant change in the obtained E_n was found anymore.

Calculation of the Matrix Elements

In order to carry out the numerical calculations, all matrix elements have to be calculated in the phase space. Combining (D.2) with (2.36) and $M = C(\Phi_0/2\pi)^2$, we find that the

---

[1] MathWorks, 3 Apple Hill Drive Natick, MA 01760-2098, USA

Hamiltonian of the phase qubit is given by

$$\hat{H} = \frac{(2e)^2}{2C}\hat{n}^2 + E_J\left(1 - \cos\hat{\varphi} + \frac{(\hat{\varphi} - 2\pi\Phi_{\text{ext}}/\Phi_0)^2}{2\beta_{L,\text{qb}}}\right),. \tag{D.4}$$

Here, the momentum operator has been replaced by the number of charge operator $\hat{n}$ according to $\hat{p} = \hbar\hat{n}$. Now, the canonical commutation relation is given by $[\hat{\varphi},\hat{n}] = i$. In order to diagonalize this Hamiltonian and hence obtain the energy eigenvalues E_n, its matrix elements have to be calculated in the discretized phase basis $|\varphi_k\rangle$ of dimension N. For the terms in (D.4) containing a function of $\hat{\varphi}$, this is trivial and yields:

$$\langle\varphi_{k'}|\hat{\varphi}|\varphi_k\rangle = \varphi\,\delta_{kk'} \tag{D.5}$$
$$\langle\varphi_{k'}|\cos\hat{\varphi}|\varphi_k\rangle = \cos\varphi\,\delta_{kk'} \tag{D.6}$$
$$\langle\varphi_{k'}|\hat{\varphi}^2|\varphi_k\rangle = \varphi^2\,\delta_{kk'} \tag{D.7}$$

where $\delta_{kk'}$ is the Kronecker delta. Determination of the charge term containing $\hat{n}^2$ is more complicated. First a transformation from phase to charge space needs to be carried out. For canonically conjugate, discrete variables, this is done by the discrete Fourier transform

$$|\varphi_k\rangle = \frac{1}{\sqrt{N}}\sum_{n=0}^{N-1} e^{in\varphi_k}|n\rangle\,. \tag{D.8}$$

Consequently, the matrix element $\hat{n}^2$ can be calculated as

$$\langle\varphi_{k'}|\hat{n}^2|\varphi_k\rangle = \frac{1}{N}\sum_{n,n'}^{N-1} n^2 e^{i(n\varphi_k - n'\varphi_{k'})}\langle n'|n\rangle = \frac{1}{N}\sum_{n=0}^{N-1} n^2 e^{in(\varphi_k - \varphi_{k'})}\,. \tag{D.9}$$

This expression has to be evaluated separately for the cases $k = k'$ and $k \neq k'$. For $k = k'$, we obtain

$$\langle\varphi_k|\hat{n}^2|\varphi_k\rangle = \frac{1}{N}\sum_{n=0}^{N-1} n^2 = \frac{(2N-1)(N-1)}{6}\,. \tag{D.10}$$

In the case $k \neq k'$, we have to define some auxiliary functions. First, we note that for a regularly discretized phase space $\varphi_k = k/N \cdot 2\pi$, so that with the definition

$$\alpha = i(\varphi_k - \varphi_{k'}) \tag{D.11}$$

we can write

$$e^{\alpha N} = e^{i2\pi(k-k')} = 1\,. \tag{D.12}$$

Second, we write the right side of (D.9) as the function $f(\lambda)$ at $\lambda = 1$:

$$f(\lambda) = \frac{1}{N}\sum_{n=0}^{N-1} n^2 e^{n\alpha\lambda} = \frac{1}{N\alpha^2}\frac{\partial^2}{\partial\lambda^2}\sum_{n=0}^{N-1}\left(e^{\alpha\lambda}\right)^n = \frac{1}{N\alpha^2}\frac{\partial^2}{\partial\lambda^2}\frac{1 - e^{\alpha\lambda N}}{1 - e^{\alpha\lambda}}\,. \tag{D.13}$$

The function $f(\lambda)$ can be calculated as

$$f(\lambda) = \frac{1}{N\alpha^2}\left\{-\frac{2N\alpha^2 e^{\alpha\lambda(N+1)}}{\left(1 - e^{\alpha\lambda}\right)^2} - \frac{N^2\alpha^2 e^{N\alpha\lambda}}{1 - e^{\alpha\lambda}} + \left(1 - e^{N\alpha\lambda}\right)\left[\frac{2\alpha^2 e^{2\alpha\lambda}}{\left(1 - e^{\alpha\lambda}\right)^3} + \frac{\alpha^2 e^{\alpha\lambda}}{\left(1 - e^{\alpha\lambda}\right)^2}\right]\right\}\,. \tag{D.14}$$

By evaluating this expression at $\lambda = 1$, using (D.12) and substituting (D.11) back in, we obtain the matrix element of $\hat{n}^2$ for $k \neq k'$:

$$\langle \varphi_{k'} | \hat{n}^2 | \varphi_k \rangle = f(\lambda)|_{\lambda=1} = \frac{(N-2)e^{i(\varphi_k - \varphi_{k'})} - N}{\left(1 - e^{i(\varphi_k - \varphi_{k'})}\right)^2} \, . \tag{D.15}$$

List of Figures

List of Tables

List of Own Publications

[KBLS11]   Ch. Kaiser, T. Bauch, F. Lombardi and M. Siegel. *Quantum Phase Dynamics in an LC shunted Josephson Junction*. Journal of Applied Physics, **109**, 093915 (2011).

[KMI$^+$11]   Ch. Kaiser, J. M. Meckbach, K. Ilin, J. Lisenfeld, R. Schäfer, A. V. Ustinov and M. Siegel. *Aluminum hard mask technique for the fabrication of high quality submicron Nb/Al-AlO$_x$/Nb Josephson junctions*. Superconductor Science and Technology, **24**, 035005 (2011).

[KSS11]   Christoph Kaiser, Roland Schäfer and Michael Siegel. *Dependence of the Macroscopic Quantum Tunneling Rate on Josephson Junction Area*. submitted to Physical Review B (2011). http://arxiv.org/abs/1011.4241.

[KSW$^+$10]   Ch. Kaiser, S. T. Skacel, S. Wünsch, R. Dolata, B. Mackrodt, A. Zorin and M. Siegel. *Measurement of dielectric losses in amorphous thin films at gigahertz frequencies using superconducting resonators*. Superconductor Science and Technology, **23**, 075008 (2010).

[KWC$^+$09]   Ch. Kaiser, G. Weiss, T. W. Cornelius, M. E. Toimil-Molares and R. Neumann. *Low temperature magnetoresistance measurements on bismuth nanowire arrays*. Journal of Physics: Condensed Matter, **21**, 205301 (2009).

[SKW$^+$11]   S. T. Skacel, Ch. Kaiser, S. Wünsch, H. Rotzinger, A. Lukashenko, M. Siegel and A. V. Ustinov. *Probing the TLS Density of States in SiO using Superconducting Lumped Element Resonators*. in preparation for Physical Review Letters (2011).

Supervised Student Theses

[Ant09] Jochen Antes. *Entwurf und Konstruktion einer Steuerung für die kontrollierte Oxidation von Aluminium.* Studienarbeit, Institut für Mikro- und Nanoelektronische Systeme, Universität Karlsruhe (TH) (2009).

[Bau10] Max Bauer. *Herstellung von Nano-Brücken mittels Elektronenstrahllithographie.* Studienarbeit, Institut für Mikro- und Nanoelektronische Systeme, Karlsruher Institut für Technologie (KIT) (2010).

[Bru08] Daniel Bruch. *Optimierung eines Printed Circuit Boards für Qubit-Experimente.* Studienarbeit, Institut für Mikro- und Nanoelektronische Systeme, Universität Karlsruhe (TH) (2008).

[Goe10] Christian Goebel. *Entwurf, Herstellung und Messung von DC-SQUIDs.* Studienarbeit, Institut für Mikro- und Nanoelektronische Systeme, Karlsruher Institut für Technologie (KIT) (2010).

[Gra07] Andreas Graf. *Optimierung eines Schwingkreises zum Auslesen eines Quantenbits.* Studienarbeit, Institut für Mikro- und Nanoelektronische Systeme, Universität Karlsruhe (TH) (2007).

[Mec08] Johannes Maximilian Meckbach. *Fabrication and experimental study of fractional flux quanta in long Josephson Junctions.* Studienarbeit, Institut für Mikro- und Nanoelektronische Systeme, Universität Karlsruhe (TH) (2008).

[Mec09] Johannes Maximilian Meckbach. *Entwicklung langer Josephson-Kontakte mit sub-μm Strom-Injektoren.* Diplomarbeit, Institut für Mikro- und Nanoelektronische Systeme, Universität Karlsruhe (TH) (2009).

[Mer10] Michael Merker. *Bestimmung der supraleitenden Kohärenzlänge in Niobfilmen und Design von NanoSQUIDs.* Studienarbeit, Institut für Mikro- und Nanoelektronische Systeme, Karlsruher Institut für Technologie (KIT) (2010).

[Rez11] Maher Rezem. *Optimierung von Ätzprozessen zur Mikro- und Nanostrukturierung.* Studienarbeit, Institut für Mikro- und Nanoelektronische Systeme, Karlsruher Institut für Technologie (KIT) (2011).

[Ska10] Sebastian Skacel. *Investigation of dielectric losses in thin-film resonators.* Diplomarbeit, Institut für Mikro- und Nanoelektronische Systeme and Physikalisches Institut, Karlsruher Institut für Technologie (KIT) (2010).

Bibliography

[1] B. D. Josephson. *Possible New Effects in Superconductive Tunnelling.* Physics Letters, **1**, 251 (1962).

[2] K. K. Likharev and V. K. Semenov. *RSFQ Logic/Memory Family: A New Josephson-Junction Technology for Sub-Terahertz-Clock-Frequency Digital Systems.* IEEE Transactions on Applied Superconductivity, **1**, 3 (1991).

[3] J. Niemeyer, J. H. Hinken and R. L. Kautz. *Microwave-induced constant-voltage steps at one volt from a series array of Josephson junctions.* Applied Physics Letters, **45**, 478 (1984).

[4] S. P. Benz and C. A. Hamilton. *A pulse-driven programmable Josephson voltage standard.* Applied Physics Letters, **68**, 3171 (1996).

[5] M. Darula, T. Doderer and S. Beuven. *Millimetre and sub-mm wavelength radiation sources based on discrete Josephson junction arrays.* Superconductor Science and Technology, **12**, R1 (1999).

[6] T. Phillips, D. Woody, G. Dolan, R. Miller and R. Linke. *Dayem-Martin (SIS tunnel junction) mixers for low noise heterodyne receivers.* IEEE Transactions on Magnetics, **17**, 684 (1981).

[7] A. M. Klushin, C. Weber, M. Siegel, S. I. Borovitskii, R. K. Starodubrovskii, K. Y. Platov, M. Y. Kupriyanov and R. Semeradk. *Integrated microwave circuits for digital-to-analogue converters based on high-temperature superconducting Josephson junction arrays.* Superconductor Science and Technology, **12**, 704 (1999).

[8] J. Clarke, W. M. Goubau and M. B. Ketchen. *Tunnel Junction dc SQUID: Fabrication, Operation, and Performance.* Journal of Low Temperature Physics, **25**, 99 (1976).

[9] E. Schrödinger. *Die gegenwärtige Situation in der Quantenmechanik.* Naturwissenschaften, **23**, 807 (1935).

[10] R. F. Voss and R. A. Webb. *Macroscopic Quantum Tunneling in 1-μm Nb Josephson Junctions.* Physical Review Letters, **47**, 265 (1981).

[11] A. J. Leggett. *Quantum Mechanics versus Macroscopic Realism: Is the Flux There when Nobody Looks?* Physical Review Letters, **54**, 857 (1985).

[12] J. R. Friedman, V. Patel, W. Chen, S. K. Tolpygo and J. E. Lukens. *Quantum superposition of distinct macroscopic states.* Nature, **406**, 43 (2000).

[13] Y. Nakamura, Y. A. Pashkin and J. S. Tsai. *Coherent control of macroscopic quantum states in a single-Cooper-pair box.* Nature, **398**, 786 (1999).

[14] J. M. Martinis, S. Nam, J. Aumentado and C. Urbina. *Rabi Oscillations in a Large Josephson-Junction Qubit.* Physical Review Letters, **89**, 117901 (2002).

[15] I. Chiorescu, Y. Nakamura, C. J. P. M. Harmans and J. E. Mooij. *Coherent Quantum Dynamics of a Superconducting Flux Qubit.* Science, **299**, 1869 (2003).

[16] D. Vion, A. Aassime, A. Cottet, P. Joyez, H. Pothier, C. Urbina, D. Esteve and M. H. Devoret. *Manipulating the Quantum State of an Electrical Circuit.* Science, **296**, 886 (2002).

[17] D. P. DiVincenzo. *The Physical Implementation of Quantum Computation.* Fortschritte der Physik, **48**, 771 (2000).

[18] R. P. Feynman. *Simulating Physics with Computers.* International Journal of Theoretical Physics, **21**, 467 (1982).

[19] P. W. Shor. *Polynomial-Time Algorithms for Prime Factorization and Discrete Logarithms on a Quantum Computer.* SIAM Journal on Computing, **26**, 1484 (1997).

[20] L. K. Grover. *A fast quantum mechanical algorithm for database search.* Proceedings, 28th Annual ACM Symposium on the Theory of Computing (STOC), **212** (1996). Available at arXiv:quant-ph/9605043v3.

[21] J. Lisenfeld, A. Lukashenko, M. Ansmann, J. M. Martinis and A. V. Ustinov. *Temperature Dependence of Coherent Oscillations in Josephson Phase Qubits.* Physical Review Letters, **99**, 170504 (2007).

[22] H. Paik, S. K. Dutta, R. M. Lewis, T. A. Palomaki, B. K. Cooper, R. C. Ramos, H. Xu, A. J. Dragt, J. R. Anderson, C. J. Lobb and F. C. Wellstood. *Decoherence in dc SQUID phase qubits.* Physical Review B, **77**, 214510 (2008).

[23] J. M. Martinis, K. B. Cooper, R. McDermott, M. Steffen, M. Ansmann, K. D. Osborn, K. Cicak, S. Oh, D. P. Pappas, R. W. Simmonds and C. C. Yu. *Decoherence in Josephson Qubits from Dielectric Loss.* Physical Review Letters, **95**, 210503 (2005).

[24] M. Tinkham. *Introduction to Superconductivity.* Second Edition (Dover Publications, Inc., Mineola, New York, 1996).

[25] A. Barone and G. Paterno. *Physics and Applications of the Josephson Effect.* First Edition (John Wiley and Sons, New York, 1982).

[26] W. Buckel and R. Kleiner. *Superconductivity - Fundamentals and Applications.* Second Edition (WILEY-VCH Verlag, Weinheim, 2004).

[27] W. C. Stewart. *Current-Voltage Characteristics of Josephson Junctions.* Applied Physics Letters, **12**, 277 (1968).

[28] D. E. McCumber. *Effect of ac Impedance on dc Voltage-Current Characteristics of Superconductor Weak-Link Junctions*. Journal of Applied Physics, **39**, 3113 (1968).

[29] M. Weihnacht. *Influence of Film Thickness on D. C. Josephson Current*. physica status solidi (b), **32**, K169 (1969).

[30] H. A. Kramers. *Brownian Motion in a Field of Force and the Diffusion Model of Chemical Reactions*. Physica (Utrecht), **7**, 284 (1940).

[31] P. Hänggi, P. Talkner and M. Borkovec. *Reaction-rate theory: fifty years after Kramers*. Reviews of Modern Physics, **62**, 251 (1990).

[32] A. O. Caldeira and A. J. Leggett. *Quantum Tunnelling in a Dissipative System*. Annals of Physics (N.Y.), **149**, 374 (1983).

[33] J. M. Martinis, M. H. Devoret and J. Clarke. *Experimental tests for the quantum behavior of a macroscopic degree of freedom: The phase difference across a Josephson junction*. Physical Review B, **35**, 4682 (1987).

[34] H. Grabert, P. Olschowski and U. Weiss. *Quantum decay rates for dissipative systems at finite temperatures*. Physical Review B, **36**, 1931 (1987).

[35] E. Freidkin, P. S. Riseborough and P. Hänggi. *The influence of dissipation on the quantal transition state tunneling rate*. Journal of Physics C: Solid State Physics, **21**, 1543 (1988).

[36] I. Affleck. *Quantum-Statistical Metastability*. Physical Review Letters, **46**, 388 (1981).

[37] E. Freidkin, P. S. Riseborough and P. Hänggi. *Quantum tunneling at low temperatures: results for weak damping*. Zeitschrift für Physik B – Condensed Matter, **64**, 237 (1986).

[38] E. Freidkin, P. S. Riseborough and P. Hänggi. *Addendum*. Zeitschrift für Physik B – Condensed Matter, **67**, 271 (1987).

[39] A. Wallraff, T. Duty, A. Lukashenko and A. V. Ustinov. *Multiphoton Transitions between Energy Levels in a Current-Biased Josephson Tunnel Junction*. Physical Review Letters, **90**, 037003 (2003).

[40] T. A. Palomaki, S. K. Dutta, R. M. Lewis, H. Paik, K. Mitra, B. K. Cooper, A. J. Przybysz, A. J. Dragt, J. R. Anderson, C. J. Lobb and F. C. Wellstood. *Pulse Current Measurements and Rabi Oscillations in a dc SQUID Phase Qubit*. IEEE Transactions on Applied Superconductivity, **17**, 162 (2007).

[41] S. K. Dutta, F. W. Strauch, R. M. Lewis, K. Mitra, H. Paik, T. A. Palomaki, E. Tiesinga, J. R. Anderson, A. J. Dragt, C. J. Lobb and F. C. Wellstood. *Multilevel effects in the Rabi oscillations of a Josephson phase qubit*. Physical Review B, **78**, 104510 (2008).

[42] R. W. Simmonds, K. M. Lang, D. A. Hite, S. Nam, D. P. Pappas and J. M. Martinis. *Decoherence in Josephson Phase Qubits from Junction Resonators*. Physical Review Letters, **93**, 077003 (2004).

[43] J. P. Carbotte. *Properties of boson-exchange superconductors*. Reviews of Modern Physics, **62**, 1027 (1990).

[44] D. R. Tilley and J. Tilley. *Superfluidity and Superconductivity* (Adam Hilger Ltd, Bristol, 1990).

[45] J. Hinken. *Supraleiter-Elektronik* (Springer-Verlag, Berlin, 1988).

[46] T. van Duzer and C. W. Turner. *Principles of superconductive devices and circuits* (North-Holland, New York, 1981).

[47] W. H. Henkels and C. J. Kircher. *Penetration Depth Measurements on Type II Superconducting Films*. IEEE Transactions on Magnetics, **13**, 63 (1977).

[48] J. M. Murduck, J. Porter, W. Dozier, R. Sandell, J. Burch, J. Bulman, C. Dang, L. Lee, H. Chan, R. W. Simon and A. H. Silver. *Niobium Trilayer Process for Superconducting Circuits*. IEEE Transactions on Magnetics, **25**, 1139 (1989).

[49] M. Kusunoki, H. Yamamori, A. Fujimaki, Y. Takai and H. Hayakawa. *High-Quality Nb/AlO_x-Al/Nb Josephson Junctions with Gap Voltage of 2.95 mV*. Japanese Journal of Applied Physics, **32**, L1609 (1993).

[50] V. Ambegaokar and A. Baratoff. *Tunneling between superconductors*. Physical Review Letters, **10**, 486 (1963).

[51] T. Imamura, T. Shiota and S. Hasuo. *Fabrication of High Quality Nb/AlO_x-Al/Nb Josephson Junctions: I - Sputtered Nb Films for Junction Electrodes*. IEEE Transactions on Applied Superconductivity, **2**, 1 (1992).

[52] C. J. Gorter and H. B. G. Casimir. Physik. Z., **35**, 963 (1934).

[53] C. J. Gorter and H. B. G. Casimir. Z. Physik, **15**, 539 (1934).

[54] B. N. Taylor and E. Burstein. *Excess Currents in Electron Tunneling Between Superconductors*. Physical Review Letters, **10**, 14 (1963).

[55] C. J. Adkins. *Multi-Particle Tunneling between Superconductors*. Review of Modern Physics, **36**, 211 (1964).

[56] N. R. Werthamer. *Nonlinear Self-Coupling of Josephson Radiation in Superconducting Tunnel Junctions*. Physical Review, **147**, 255 (1966).

[57] M. Maezawa, M. Aoyagi, H. Nakagawa, I. Kurosawa and S. Takada. *Observation of Josephson self-coupling in Nb/AlO_x/Nb tunnel junctions*. Physical Review B, **50**, 9664 (1994).

[58] J. R. Schrieffer and J. W. Wilkins. *Two-Particle Tunneling Processes Between Superconductors.* Physical Review Letters, **10**, 17 (1963).

[59] M. Octavio, M. Tinkham, G. E. Blonder and T. M. Klapwijk. *Subharmonic energy-gap structure in superconducting constrictions.* Physical Review B, **27**, 6739 (1983).

[60] G. B. Arnold. *Superconducting Tunneling without the Tunneling Hamiltonian. II. Subgap Harmonic Structure.* Journal of Low Temperature Physics, **68**, 1 (1987).

[61] A. F. Andreev. *Thermal conductivity of the intermediate state of superconductors.* Soviet Physics - Journal of Experimental and Theoretical Physics, **19**, 1228 (1964).

[62] M. Iansiti, M. Tinkham, A. T. Johnson, W. F. Smith and C. J. Lobb. *Charging effects and quantum properties of small superconducting tunnel junctions.* Physical Review B, **39**, 6465 (1989).

[63] A. T. Johnson, C. J. Lobb and M. Tinkham. *Effect of Leads and Energy Gap upon the Retrapping Current of Josephson Junctions.* Physical Review Letters, **65**, 1263 (1990).

[64] F. P. Milliken, R. H. Koch, J. R. Kirtley and J. R. Rozen. *The subgap current in Nb/AlOx/Nb tunnel junctions.* Applied Physics Letters, **85**, 5941 (2004).

[65] M. A. Gubrud, M. Ejrnaes, A. J. Berkley, R. C. R. Jr., I. Jin, J. R. Anderson, A. J. Dragt, C. J. Lobb and F. C. Wellstood. *Sub-Gap Leakage in Nb/AlO$_x$/Nb andAl/AlO$_x$/Al Josephson Junctions.* IEEE Transactions on Applied Superconductivity, **11**, 1002 (2001).

[66] M. F. Bocko, A. M. Herr and M. J. Feldman. *Prospects for quantum coherent computation using superconducting electronics.* IEEE Transactions on Applied Superconductivity, **7**, 3638 (1997).

[67] T. Imamura and S. Hasuo. *Cross-sectional transmission electron microscopy observation of Nb/AlO$_x$-Al/Nb Josephson junctions.* Applied Physics Letters, **58**, 645 (1991).

[68] H. Kohlstedt, F. König, P. Henne, N. Thyssen and P. Caputo. *The role of surface roughness in the fabrication of stacked Nb/Al-AlO$_x$/Nb tunnel junctions.* Journal of Applied Physics, **80**, 5512 (1996).

[69] T. Imamura and S. Hasuo. *Fabrication of High Quality Nb/AlO$_x$-Al/Nb Josephson Junctions: II - Deposition of Thin Al Layers on Nb Films.* IEEE Transactions on Applied Superconductivity, **2**, 84 (1992).

[70] P. Singer. *Film Stress and How to Measure It.* Semiconductor International, **15**, 54 (1992).

[71] R. E. Miller, W. H. Mallison, A. W. Kleinsasser, K. A. Delin and E. M. Macedo. *Niobium trilayer Josephson tunnel junctions with ultrahigh critical current densities.* Applied Physics Letters, **63**, 10 (1993).

[72] L. P. S. Lee, E. R. Arambula, G. Hanaya, C. Dang, R. Sandell and H. Chan. *RHEA (Resist-Hardened Etch and Anodization) Process for Fine-Geometry Josephson Junction Fabrication.* IEEE Transactions on Magnetics, **27**, 3133 (1991).

[73] L. Fritzsch, H. Elsner, M. Schubert and H.-G. Meyer. *SNS and SIS Josephson junctions with dimensions down to the submicron region prepared by a unified technology.* Superconductor Science and Technology, **12**, 880 (1999).

[74] M. B. Ketchen, D. Pearson, A. W. Kleinsasser, C.-K. Hu, M. Smyth, J. Logan, K. Stawiasz, E. Baran, M. Jaso, T. Ross, K. Petrillo, M. Manny, S. Basavaiah, S. Brodsky, S. B. Kaplan, W. J. Gallagher and M. Bhushan. *Sub-μm, planarized, Nb-AlO$_x$-Nb Josephson process for 125 mm wafers developed in partnership with Si technology.* Applied Physics Letters, **59**, 2609 (1991).

[75] A. B. Pavolotsky, T. Weimann, H. Scherer, J. Niemeyer, A. B. Zorin and V. A. Krupenin. *Novel Method for Fabricating Deep Submicron Nb/AlO$_x$/Nb Tunnel Junctions Based on Spin-on Glass Planarization.* IEEE Transactions on Applied Superconductivity, **9**, 3251 (1999).

[76] I. Péron, P. Pasturel and K.-F. Schuster. *Fabrication of SIS Junctions for Space Borne Submillimeter Wave Mixers using Negative Resist E-beam Lithography.* IEEE Transactions on Applied Superconductivity, **11**, 377 (2001).

[77] W. Chen, V. Patel and J. E. Lukens. *Fabrication of high-quality Josephson junctions for quantum computation using a self-aligned process.* Microelectronic Engineering, **73-74**, 767 (2004).

[78] T. Imamura and S. Hasuo. *A submicrometer Nb/AlO$_x$/Nb Josephson junction.* Journal of Applied Physics, **64**, 1586 (1988).

[79] Z. Bao, M. Bhushan, S. Han and J. E. Lukens. *Fabrication of High Quality, Deep-Submicron Nb/AlO$_x$/Nb Josephson Junctions Using Chemical Mechanical Polishing.* IEEE Transactions on Applied Superconductivity, **5**, 2731 (1995).

[80] R. Dolata, T. Weimann, H.-J. Scherer and J. Niemeyer. *Sub pm Nb/AlO$_x$/Nb Josephson Junctions Fabricated by Anodization Techniques.* IEEE Transactions on Applied Superconductivity, **9**, 3255 (1999).

[81] R. Dolata, H. Scherer, A. B. Zorin and J. Niemeyer. *Single-charge devices with ultrasmall Nb/AlO$_x$/Nb trilayer Josephson junctions.* Journal of Applied Physics, **97**, 54501 (2005).

[82] S. K. Tolpygo, D. Amparo, A. Kirichenko and D. Yohannes. *Plasma process-induced damage to Josephson tunnel junctions in superconducting integrated circuits.* Superconductor Science and Technology, **20**, S341 (2007).

[83] H. Yamamori, T. Yamada, H. Sasaki and A. Shoji. *A 10 V programmable Josephson voltage standard circuit with a maximum output voltage of 20 V.* Superconductor Science and Technology, **21**, 105007 (2008).

[84] R. McDermott. *Materials Origins of Decoherence in Superconducting Qubits.* IEEE Transactions on Applied Superconductivity, **19**, 2 (2009).

[85] X. Meng and T. V. Duzer. *Light-Anodization Process for High-J_c Micron and Submicron Superconducting Junction and Integrated Circuit Fabrication.* IEEE Transactions on Applied Superconductivity, **13**, 91 (2003).

[86] W. Heni. *Untersuchung der Isolationseigenschaften und der dielektrischen Verluste in SiO_2.* Bachelor thesis, Institut für Mikro- und Nanoelektronische Systeme, Karlsruher Institut für Technologie (KIT) (2010).

[87] G. Hammer, S. Wuensch, M. Roesch, K. Ilin, E. Crocoll and M. Siegel. *Superconducting coplanar waveguide resonators for detector applications.* Superconductor Science and Technology, **20**, S408 (2007).

[88] P. Macha, S. H. W. van der Ploeg, G. Oelsner, E. Il'ichev, H.-G. Meyer, S. Wünsch and M. Siegel. *Losses in coplanar waveguide resonators at millikelvin temperatures.* Applied Physics Letters, **96**, 062503 (2010).

[89] K. K. Likharev. *Superconducting weak links.* Reviews of Modern Physics, **51**, 101 (1979).

[90] S. Nagasawa, K. Hinode, T. Satoh, H. Akaike, Y. Kitagawa and M. Hidaka. *Development of advanced Nb process for SFQ circuits.* Physica C, **412**, 1429 (2004).

[91] J. M. Meckbach. Internal report, Institut für Mikro- und Nanoelektronische Systeme (IMS), Karlsruher Institut für Technologie (KIT) (2010).

[92] J. M. Martinis. *Superconducting phase qubits.* Quantum Information Processing, **8**, 81 (2009).

[93] M. Tanaka, T. Kondo, N. Nakajima, T. Kawamoto, Y. Yamanashi, Y. Kamiya, A. Akimoto, A. Fujimaki, H. Hayakawa, N. Yoshikawa, H. Terai, Y. Hashimoto and S. Yorozu. *Demonstration of a single-flux-quantum microprocessor using passive transmission lines.* IEEE Transactions on Applied Superconductivity, **15**, 400 (2005).

[94] A. D. O'Connell, M. Ansmann, R. C. Bialczak, M. Hofheinz, N. Katz, E. Lucero, C. McKenney, M. Neeley, H. Wang, E. M. Weig, A. N. Cleland and J. M. Martinis. *Microwave dielectric loss at single photon energies and millikelvin temperatures.* Applied Physics Letters, **92**, 112903 (2008).

[95] J. Gao, M. Daal, A. .Vayonakis, S. Kumar, J. Zmuidzinas, B. Sadoulet, B. A. Mazin, P. K. Day and H. G. Leduc. *Experimental evidence for a surface distribution of two-level systems in superconducting lithographed microwave resonators.* Applied Physics Letters, **92**, 152505 (2008).

[96] W. Chen, D. A. Bennett, V. Patel and J. E. Lukens. *Microwave dielectric loss at single photon energies and millikelvin temperatures.* Superconductor Science and Technology, **21**, 075013 (2008).

[97] A. K. Jonscher. *Dielectric relaxation in solids.* Journal of Physics D: Applied Physics, **32**, R57 (1999).

[98] A. K. Jonscher. *The 'universal' dielectric response.* Nature, **267**, 673 (1977).

[99] K. L. Ngai, A. K. Jonscher and C. T. White. *On the origin of the universal dielectric response in condensed matter.* Nature, **277**, 185 (1979).

[100] M. Hein. *High-Temperature-Superconductor Thin Films at Microwave Frequencies* (Springer Verlag, Berlin Heidelberg, 1999).

[101] A. P. S. Khanna and Y. Garault. *Determination of Loaded, Unloaded, and External Quality Factors of a Dielectric Resonator Coupled to a Microstrip Line.* IEEE Transactions on Microwave Theory and Techniques, **31**, 261 (1983).

[102] W. A. Phillips. *Two-level states in glasses.* Reports on Progress in Physics, **50**, 1657 (1987).

[103] A. Shnirman, G. Schön, I. Martin and Y. Makhlin. *Low- and High-Frequency Noise from Coherent Two-Level Systems.* Physical Review Letters, **94**, 127002 (2005).

[104] T. A. Fulton and L. N. Dunkleberger. *Lifetime of the zero-voltage state in Josephson tunnel junctions.* Physical Review B, **9**, 4760 (1974).

[105] G. L. Ingold and Y. V. Nazarov. *Charge Tunneling Rates in Ultrasmall Junctions.* In H. Grabert and M. H. Devoret (Editors), *Single Charge Tunneling*, volume 294 of *NATO ASI Series B*, 21 (Plenum Press, New York, 1992). also available at arXiv:cond-mat/0508728v1.

[106] M. Büttiker, E. P. Harris and R. Landauer. *Thermal activation in extremely underdamped Josephson-junction circuits.* Physical Review B, **28**, 1268 (1983).

[107] H. Grabert and U. Weiss. *Crossover from Thermal Hopping to Quantum Tunneling.* Physical Review Letters, **53**, 1787 (1984).

[108] A. Wallraff, A. Lukashenko, C. Coqui, A. Kemp, T. Duty and A. V. Ustinov. *Switching current measurements of large area Josephson tunnel junctions.* Review of Scientific Instruments, **74**, 3740 (2003).

[109] Y. N. Ovchinnikov, A. Barone and A. A. Varlamov. *Effect of magnetic field on macroscopic quantum tunneling escape time in small Josephson junctions.* Physical Review B, **78**, 054521 (2008).

[110] Y. N. Ovchinnikov, A. Barone and A. A. Varlamov. *Macroscopic Quantum Tunneling in "Small" Josephson Junctions in a Magnetic Field.* Physical Review B, **99**, 037004 (2007).

[111] K. Fedorov. private communication (2010).

[112] T. Bauch, F. Lombardi, F. Tafuri, A. Barone, G. Rotoli, P. Delsing and T. Claeson. *Macroscopic Quantum Tunneling in d-Wave $YBa_2Cu_3O_{7-\delta}$ Josephson Junctions.* Physical Review Letters, **94**, 087003 (2005).

[113] T. Bauch, T. Lindström, F. Tafuri, G. Rotoli, P. Delsing, T. Claeson and F. Lombardi. *Quantum Dynamics of a d-Wave Josephson Junction.* Science, **311**, 57 (2006).

[114] F. Lombardi, T. Bauch, G. Rotoli, T. Lindström, J. Johannson, K. Cedergren, F. Tafuri and T. Claeson. *Dynamics of a LC Shunted $YBa_2Cu_3O_{7-\delta}$ Josephson Junction.* IEEE Transactions on Applied Superconductivity, **17**, 653 (2007).

[115] G. Rotoli, T. Bauch, T. Lindstrom, D. Stornaiuolo, F. Tafuri and F. Lombardi. *Classical resonant activation of a Josephson junction embedded in an LC circuit.* Physical Review B, **75**, 144501 (2007).

[116] L. B. Ioffe, V. B. Geshkenbein, M. V. Feigelman, A. L. Fauchère and G. Blatter. *Environmentally decoupled sds-wave Josephson junctions for quantum computing.* Nature, **398**, 679 (1999).

[117] M. Steffen, M. Ansmann, R. McDermott, N. Katz, R. C. Bialczak, E. Lucero, M. Neeley, E. M. Weig, A. N. Cleland and J. M. Martinis. *State tomography of capacitively shunted phase qubits with high fidelity.* Physical Review Letters, **97**, 050502 (2006).

[118] A. Lupascu, C. J. M. Verwijs, R. N. Schouten, C. J. P. M. Harmans and J. E. Mooij. *Nondestructive readout for a superconducting flux qubit.* Physical Review Letters, **93**, 177006 (2004).

[119] J. Clarke and A. Braginski. *The SQUID Handbook, Vol. 1: Fundamentals and technology of SQUIDs and SQUID systems* (WILEY-VCH, Weinheim, 2004).

[120] J. Lisenfeld. *Experiments on Superconducting Josephson Phase Quantum Bits.* Ph.D. thesis, Naturwissenschaftliche Fakultäten der Friedrich-Alexander-Universität Erlangen-Nürnberg (2007).

[121] F. Meier and D. Loss. *Reduced visibility of Rabi oscillations in superconducting qubits.* Physical Review B, **71**, 094519 (2005).

[122] A. K. Feofanov, V. A. Oboznov, V. V. Bol'ginov, J. Lisenfeld, S. Poletto, V. V. Ryazanov, A. N. Rossolenko, M. Khabipov, D. Balashov, A. B. Zorin, P. N. Dmitriev, V. P. Koshelets and A. V. Ustinov. *Implementation of superconductor-ferromagnet-superconductor π-shifters in superconducting digital and quantum circuits.* Nature Physics, **6**, 593 (2010).

[123] D. A. Bennett, L. Longobardi, V. Patel, W. Chen and J. E. Lukens. *rf-SQUID qubit readout using a fast flux pulse.* Superconductor Science and Technology, **20**, S445 (2007).

[124] E. Hoskinson, F. Lecocq, N. Didier, A. Fay, F. W. J. Hekking, W. Guichard, O. Buisson, R. Dolata, B. Mackrodt and A. B. Zorin. *Quantum Dynamics in a Camelback Potential of a dc SQUID*. Physical Review Letters, **102**, 097004 (2009).

[125] J. Lisenfeld. private communication (2010).

[126] G. E. P. Box and K. B. Wilson. Journal of the Royal Statistical Society: Series B (Statistical Methodology), **13**, 1 (1951).

[127] M. W. Jenkins, M. T. Mocella, K. D. Allen and H. H. Sawin. *The Modelling of Plasma Etching Processes Using Response Surface Methodology*. Solid State Technology, **April**, 175 (1986).

Nomenclature

$\lvert 0 \rangle$	Ground state of a phase qubit
$\lvert 1 \rangle$	Excited state of a phase qubit
A	Area, e.g. of a Josephson junction or a capacitor [μm^2]
$\vec{A}$	Magnetic vector potential [Vs/m]
α	Auxiliary variable in calculation of phase qubit matrix elements
α_{KLD}	Thermal prefactor a_t in the Kramers very low damping limit
α_{KMD}	Thermal prefactor a_t in the Kramers moderate to high damping limit
a_q	Prefactor in the quantum tunneling rate
a_t	Thermal prefactor in the thermal escape rate
b	Exponent in fit of normal resistance to junction area
$\vec{B}$	Magnetic flux density [T]
B_{c2}	Second critical magnetic field of a superconductor [T]
β	Ratio of shell to junction capacitance $= L_s/L_{J0}$
β_C	Stewart-McCumber parameter $= \frac{2e}{\hbar} I_c R^2 C$
β_L	Screening parameter of a DC-SQUID
$\beta_{L,\mathrm{qb}}$	Screening parameter of a phase qubit
C	Capacitance, e.g. that of a Josephson junction [pF]
c	Specific capacitance of a Josephson junction [fF/μm^2]
C_c	Coupling capacitor of a resonator [fF]
χ	Complex electric susceptibility
c_i	Complex coefficients describing the superposition of the qubit states
C_J	Capacitance of a Josephson junction [pF]
C_s	Shell capacitance in an LCJJ system [pF]

C_{sh}	Shunt capacitance in a phase qubit [pF]
D	Encroachment of Nb_2O_5 during anodic oxidation [nm]
d	Length or diameter of a Josephson junction [μm]
Δ	Energy gap of a superconductor [meV], the energy needed to separate a Cooper pair is 2Δ
Δ_{01}	Level splitting of a phase qubit [$h\cdot$ GHz]
Δ_{12}	Splitting between second- and third-lowest phase qubit state [$h\cdot$ GHz]
ΔE	Energy distance of the lowest energy levels in a Josephson junction [meV]
Δf	Avoided level crossing in qubit spectrum [MHz]
δf	Bandwidth of the resonance dip [MHz]
ΔI	Current channel width of switching current histogram [μA]
ΔI_n	Amplitude of the low frequency current noise [pA]
$\Delta\Phi_{\text{sq}}$	Qubit signal strength in DC-SQUID [$m\Phi_0$]
ΔT^2	Deviation of T_{esc} from bath temperature in MQT data analysis [K^2]
ΔU	Potential barrier in a system containing a Josephson junction [meV]
d_{F}	Nb film thickness [nm]
d_i	Thickness of a dielectric layer [nm]
d_j	Coefficients in RSM modeling
$\vec{dl}$	Line integral
d_{s}	Thickness of the silicon substrate [μm]
$\vec{dS}$	Surface integral
E	Energy [J]
e	Elementary charge $= 1.6022\cdot 10^{-19}$ C
$\vec{E}$	Electric field [V/m]
E_{J}	Josephson coupling energy $= \Phi_0 I_{\text{c}}/2\pi$
ε_0	Vacuum permittivity $= 8.8542\cdot 10^{-12}$ F/m
ε_{r}	Permittivity
η	Damping coefficient in a Josephson junction $= 1/R\cdot(\Phi_0/2\pi)^2$

E_Y	Young's modulus [GPa]
F	Total flow in RIE $= F_1 + F_2$
f	Frequency [Hz]
f_{01}	Larmor frequency of a phase qubit [GHz]
F_1	CF_4 flow in RIE [sccm]
F_2	O_2 flow in RIE [sccm]
f_J	Josephson frequency $= 483.6 \frac{\text{GHz}}{\text{mV}} \cdot V$
f_{Ramsey}	Oscillation frequency of Ramsey fringes [MHz]
f_{res}	Resonance frequency of a resonator [GHz]
G	Amplification
$\tilde{G}$	Conductance [$1/\Omega$]
Γ	Escape rate from a potential well [Hz]
Γ	Offset of fit in determination of T_{esc}
γ	Normalized bias current I/I_c through a Josephson junction
γ_{cr}	Normalized bias current at the crossover to the quantum regime
Γ_ϕ	Dephasing rate in a superconducting qubit [Hz]
Γ_q	Quantum tunneling rate [Hz]
Γ_{th}	Thermal escape rate [Hz]
$\hat{H}$	Hamiltonian, e.g. for the phase qubit
H	Magnetic field [A/m]
h	Planck constant $= 6.626 \cdot 10^{-34}$ Js
$\hbar$	Reduced Planck constant $= 1.0546 \cdot 10^{-34}$ Js
H_c	Critical field of a superconductor [A/m]
I	Bias current through a system containing a Josephson junction [μA]
I_{AO}	Current in the electrolyte during anodic oxidation [μA]
I_c	Critical current of a Josephson junction [μA]
I_{c0}	Theoretical critical current of a Josephson junction in the absence of any fluctuations [μA]

I_{circ}	Circulating current in a SQUID loop [μA]
$\dot{I}$	Current ramp rate in MQT experiments [μA/s]
I_{exc}	Excess current of a Josephson junction [μA]
I_{magn}	Current through a coil creating a magnetic field [mA]
I_{MW}	Microwave amplitude [μV]
I_r	Retrapping current of a Josephson junction [μA]
I_{sw}	Switching current of a Josephson junction [μA]
$I_{sw,cr}$	Switching current in the quantum regime [μA]
j_c	Critical current density of a Josephson junction [A/cm^2]
$\vec{j}_s$	Supercurrent density in a superconductor [A/cm^2]
$j_{s,c}$	Critical current density of a superconductor [A/cm^2]
K	Bending of silicon wafers under film stress [μm]
κ	Exponent in dependence of critical current density on oxygen exposure
$\varkappa, \varkappa_i$	Coefficient(s) in formula $\Omega_R(I_{MW})$
k_B	Boltzmann's constant $= 1.3807 \cdot 10^{-23}$ J/K
L	Inductance [nH]
ℓ	Electronic mean free path [nm]
λ	Auxiliary variable in calculation of phase qubit matrix elements
λ_L	London penetration depth [nm]
λ_J	Josephson penetration depth [nm]
L_J	Josephson inductance [pH]
L_{J0}	Josephson inductance for $I \ll I_c$ [pH]
L_{qb}	Inductance of a phase qubit loop [pH]
L_s	Shell inductance in an LCJJ system [nH]
L_{scan}	Scan length along silicon wafers [mm]
L_{sq}	Inductance of a SQUID loop [pH]
M	Mass of the phase particle $= C(\Phi_0/2\pi)^2$

M_{12}	Mutual inductive coupling of phase qubit and bias coil [pH]
M_{13}	Inductive coupling between bias coil and DC-SQUID [pH]
M_{23}	Mutual inductive coupling of phase qubit and DC-SQUID [pH]
m_s	Mass of the superconducting charge carriers [kg]
μ_0	Vacuum permeability $= 4\pi \cdot 10^{-7}$ H/m
$\hat{n}$	Number of charge operator
$\mathbb{N}$	Set of all natural numbers
N	Dimension of the Hilbert space
n	A number, for example in the exponent of the universal law; an integer, for example in the context of magnetic flux quantization or MAR
$\vec{\nabla}$	Nabla operator
n_s	Cooper pair density [cm^{-3}]
ν	Poisson number
O	Oxygen content in process gas during RIE [%]
ω	Angular frequency $= 2\pi f$
ω_{MW}, f_{MW}	Frequency of microwave irradiation [GHz]
ω_p	Plasma frequency of a Josephson junction [GHz]
ω_{p0}	Plasma frequency of a Josephson junction at zero bias $= \sqrt{\frac{2eI_c}{\hbar C}}$
ω_{peak}	Peak frequency for dielectric losses $= 1/\tau$
$\omega_{\pm}$	Lower and upper plasma mode in an LCJJ system [GHz]
Ω_R	Rabi-frequency including the second excited qubit state [MHz]
ω_{Rabi}, f_{Rabi}	Rabi-frequency of a phase qubit [MHz]
$\hat{p}$	Momentum operator
P	Power, e.g. in RIE or sputtering processes [W]
p	Pressure, e.g. during RIE [mTorr]
p_{Ar}	Working pressure during Nb sputtering [mTorr]
P_{esc}	Escape probability from the potential well of a phase qubit

Φ	Magnetic flux [Wb]
ϕ	Phase of the superconducting wave function
Φ_0	Magnetic flux quantum $= h/2e = 2.0678 \cdot 10^{-15}$ Wb
Φ_1	Flux bias for qubit preparation in the shallow potential well [Wb]
φ	Gauge-invariant phase difference across a Josephson junction
Φ_2	Flux bias for qubit operation [Wb]
$\hat{\varphi}$	Josephson phase operator
Φ_{ext}	Externally applied magnetic flux [Wb]
$|\varphi_k\rangle$	Basis states in discretized phase space
$\varphi_s, \varphi_L, \varphi_J$	Phase variables in an LCJJ system
$P(I)$	Probability distribution of switching currents [1/μA]
p_i	Probability to find the qubit in state $|i\rangle$
ϖ	Polar angle in the Bloch sphere
p_{oxy}	Oxygen pressure during Al oxidation [mbar]
ψ	Superconducting wave function
$|\psi\rangle$	Arbitrary pure qubit state
Q	Quality factor of a Josephson junction $= \omega_p RC$
q	Electric charge [C]
Q_0	Intrinsic quality factor of a resonator
Q_C	Coupling quality factor of a resonator
Q_ε	Quality factor of a resonator due to dielectric losses
Q_L	Loaded quality factor of a resonator
Q_{rad}	Quality factor of a resonator due to radiation losses
Q_ρ	Quality factor of a resonator due to conductor losses
q_s	Charge of the superconducting charge carriers $= 2e$ [C]
R	Resistance, e.g. that of a Josephson junction [Ω]
R_0	Load resistor of a resonator [Ω]

R_{curv}	Radius of curvature of silicon wafers under film stress [m]
r_{dep}	Deposition rate, e.g. during Nb sputtering [nm/s]
$r_{\mathrm{e,Nb}}$	Nb etching rate with RIE [nm/min]
$r_{\mathrm{e,res}}$	Resist etching rate with RIE [nm/min]
ρ	Resistivity [Ωm]
ρ_0	Resistivity at low temperatures [Ωm]
ρ_{fit}	Prefactor in fit of normal resistance to junction area
ρ_{Ph}	Resistivity due to phonon scattering [Ωm]
R_{N}	Normal resistance of a Josephson junction [Ω]
R_{QP}	Quasiparticle resistance of a Josephson junction [Ω]
RRR	Residual resistance ratio
R_{s}	Surface impedance of a superconductor [$\mu\Omega$]
R_{sg}	Subgap resistance of a Josephson junction [Ω]
$R_{\mathrm{sg,intr}}$	Intrinsic subgap resistance of a Josephson junction [Ω]
$R_{\mathrm{sg,max}}$	Maximal subgap resistance of a Josephson junction [Ω]
R_{sh}	Shunting resistance, e.g. of a Josephson junction [Ω]
S_{12}	Transmission scattering matrix parameter
$S_{12,\mathrm{min}}$	Transmission scattering matrix parameter at resonance dip
σ	Nb film stress [GPa]
σ_{dev}	Standard deviation, e.g. for the critical current of Josephson junctions
t	Time coordinate [s]
T, T_{bath}	(Bath) temperature [K]
t_0	Dead time for RIE [s]
$\tilde{t}$	Normalized time for a Josephson junction $= 2eI_{\mathrm{c}}R/\hbar \cdot t$
T_1	Relaxation time in a superconducting qubit [ns]
T_2	Dephasing time in a superconducting qubit [ns]
t_{a}	Time of anodic oxidation [s]

$\tan\delta$	Loss tangent, e.g. for dielectric losses
τ	Relaxation time of dipoles in a dielectric [ns]
$1/\tau_{\mathrm{qp}}$	Decoherence rate from quasiparticle conductance [1/µs]
$\tau_{\mathrm{qb,TLS}}$	Characteristic time for interaction between qubit and TLS [ns]
T_{c}	Critical temperature of a superconductor [K]
T_{cr}	Crossover temperature from the thermal to the quantum regime [K]
T_{d}	Decay time of Rabi oscillations [ns]
t_{delay}	Delay time between microwave pulse and qubit read out
T_{esc}	Escape temperature in a Josephson junction [K]
θ	Azimuthal angle in the Bloch sphere
ϑ	Ratio of junction to shell capacitance $= C_{\mathrm{J}}/C_{\mathrm{s}}$
t_{MW}	Duration of microwave pulse [ns]
t_{ox}	Thickness of tunneling oxide (AlO_x) in a Josephson junction [nm]
t_{oxy}	Oxidation time of AlO_x tunneling barrier [s]
t_{sw}	Time between $I = 0$ and switching to the voltage state [s]
$U(\varphi)$	Potential of a system containing a Josephson junction [meV]
$\Upsilon_c(x)$	Integral critical current density in a Josephson junction [A/m]
V	Voltage, e.g. across a Josephson junction [V]
V_{gap}	Gap voltage of a Josephson junction [mV]
V_{m}	Characteristic voltage of a Josephson junction $= I_{\mathrm{c}}R_{\mathrm{sg}}$
w	Width of a Josephson junction [µm]
$\vec{x} = (x,y,z)$	Spatial coordinates [m]
X	Reactance $= \mathrm{Im}Z$
ξ_0	BCS coherence length [nm]
$\xi(T)$	Ginzburg-Landau coherence length [nm]
Y	Admittance [1/Ω]
Y_i	Response in RSM modeling, e.g. the Nb etching rate

Z	Complex impedance $[\Omega]$
Z_0	Impedance of the bias lines of a Josephson junction or a phase qubit $[\Omega]$
Z_{eff}	Effective impedance of the bias lines for a phase qubit $[\Omega]$
ζ	Slope of fit in determination of T_{esc}
$z_{i,j}$	Combinations of parameter values in RSM modeling, e.g. $p \cdot F$